Comparing Armed Conflicts

Comparing armed conflicts primes the production of categories that, when mobilized, can alter the trajectories of the conflicts. Considering the political dynamism of spaces of conflict and intervention, and that practitioners regularly seek out academic expertise, this book discusses the possibilities and limits of comparative approaches to understanding armed conflict and intervention.

Capacity-building experts, development workers, international organizations, and diplomats use their previous experiences and bring them into new contexts to understand and respond to their environment. Conflict actors can also make comparisons to buttress their political position in negotiations, consolidate their control over fighters, and as calls for transnational rebel solidarity. The use of such comparisons is an inherently political move and it has an impact on the production of scientific knowledge, on conflict dynamics themselves, and on the formulation and implementation of conflict management policy: comparison is inherently a practice of order-making. While there are important epistemological and methodological stakes associated with researchers engaging in comparison, there are also important productive effects connected to the research avenues taken.

The chapters in this book were originally published in the *Civil Wars*.

Bruno Charbonneau is Professor and Director of the Centre for Security and Crisis Governance (CRITIC) at the Royal Military College Saint-Jean, and the Founder and Director of the Centre FrancoPaix in Conflict Resolution of the Raoul-Dandurand Chair, Université du Québec à Montréal (UQAM), Canada.

Adam Sandor is Research Associate with the Centre FrancoPaix in Conflict Resolution of the Raoul-Dandurand Chair, Université du Québec à Montréal (UQAM), Canada.

Comparing Armed Conflicts

Edited by
Bruno Charbonneau and Adam Sandor

Routledge
Taylor & Francis Group
LONDON AND NEW YORK

First published 2022
by Routledge
2 Park Square, Milton Park, Abingdon, Oxon, OX14 4RN

and by Routledge
605 Third Avenue, New York, NY 10158

Routledge is an imprint of the Taylor & Francis Group, an informa business

British Library Cataloguing-in-Publication Data
A catalogue record for this book is available from the British Library

ISBN13: 978-1-032-01637-5 (hbk)
ISBN13: 978-1-032-01640-5 (pbk)
ISBN13: 978-1-003-17940-5 (ebk)

Typeset in Myriad Pro
by codeMantra

Publisher's Note
The publisher accepts responsibility for any inconsistencies that may have arisen during the conversion of this book from journal articles to book chapters, namely the inclusion of journal terminology.

Contents

Citation Information

The chapters in this book, except Chapter 7, were originally published in the *Civil Wars*, volume 21, issue 4 (2019). Chapter 7 was originally published in volume 22, issue 4 (2020) of the same journal. When citing this material, please use the original page numbering for each article, as follows:

Introduction
Power and Comparative Methods: Performing the Worlds of Armed Conflicts
Bruno Charbonneau and Adam Sandor
Civil Wars, volume 21, issue 4 (2019) pp. 437–447

Chapter 1
The Importance of Context When Comparing Civil Wars
William Reno
Civil Wars, volume 21, issue 4 (2019) pp. 448–467

Chapter 2
Comparing Conflict-related Sexual Violence: Expertise, Politics and Documentation
Róisín Read
Civil Wars, volume 21, issue 4 (2019) pp. 468–488

Chapter 3
Fanning the Flames or a Troubling Truth? The Politics of Comparison in the Israel-Palestine Conflict
Mandy Turner
Civil Wars, volume 21, issue 4 (2019) pp. 489–513

Chapter 4
Privileged Sphere of Comparison: Empire, Methods and Conflict Intervention
Bruno Charbonneau
Civil Wars, volume 21, issue 4 (2019) pp. 514–538

Chapter 5

The Middle East is Violence: On the Limits of Comparative Approaches to the Study of Armed Conflict
Jacob Mundy
Civil Wars, volume 21, issue 4 (2019) pp. 539–568

Chapter 6

Normative Scaling and Crisis Knowledge: The Problematic Use of Selective Analogies to Compare Conflicts
Florian P. Kühn
Civil Wars, volume 21, issue 4 (2019) pp. 569–593

Chapter 7

Sahelistan? Military Intervention and Patronage Politics in Afghanistan and Mali
Romain Malejacq and Adam Sandor
Civil Wars, volume 22, issue 4 (2020) pp. 543–566

For any permission-related enquiries please visit:
http://www.tandfonline.com/page/help/permissions

Contributors

Bruno Charbonneau International Studies, Royal Military College Saint-Jean, Richelain, QC, Canada.

Florian P. Kühn Käte Hamburger Kolleg, Centre for Global Cooperation Research, University of Duisburg-Essen, Germany.

Romain Malejacq Centre for International Conflict Analysis & Management (CICAM), Institute for Management Research, Radboud University Nijmegen, The Netherlands.

Jacob Mundy Peace and Conflict Studies, Colgate University, Hamilton, USA.

Róisín Read Humanitarian and Conflict Response Institute, The University of Manchester, UK.

William Reno Department of Political Science, Northwestern University, Evanston, USA.

Adam Sandor Centre FrancoPaix in Conflict Resolution and Peace Missions, University of Quebec in Montreal (UQÀM), Canada.

Mandy Turner Humanitarian and Conflict Response Institute, The University of Manchester, UK.

Introduction: Power and Comparative Methods: Performing the Worlds of Armed Conflicts

Bruno Charbonneau and Adam Sandor

ABSTRACT

This Special Issue emphasises how power and power relations involved in establishing limits and boundaries to define, categorise and understand the world through comparison are intimately tied to conflict and intervention practices and dynamics. Indeed, when pundits, practitioners, academics and even conflict actors compare settings of armed conflict and intervention, they are participating in an inherently political move. The most off-handed of comments connect to assemblages that enable the production of categories and concepts from which it becomes difficult to think differently. Our comparisons perform worlds of armed conflict, and international interventions more often than not reflect those performances.

Making comparisons is no innocuous act. When pundits, practitioners, and academics compare settings of armed conflict and intervention, they are participating in an inherently political move. Of course, not all such political moves are equal in their consequences. It nevertheless remains the case that even the most off-handed of comments connect to material and ideational assemblages that enable the production of hegemonic categories and concepts from which it becomes difficult to think differently. Our comparisons perform worlds of armed conflict, and international interventions more often than not reflect those performances.

This Special Issue was borne out of frustrations over repeatedly hearing commentators and analysts compare the conflict in Mali to that of Afghanistan. As the Malian conflict escalated and calls for the need to combat jihadist terrorists were amplified by geopolitical actors, new terms such as 'Sahelistan' (Laurent 2013) and 'Africanistan' (Michailof 2015) began circulating unproblematically, and apolitically – and continue to do so now. Back in 2013, even the British Prime Minister compared the conflict in Mali to the so-called existential threat of terrorism in Afghanistan (BBC 2013). Dutch,

German, Canadian, and some French and American interveners with whom we spoke during fieldwork, having worked in Afghanistan, felt very comfortable conflating the two countries' conflict dynamics. African government officials likewise pursued this discursive trend – although perhaps for more strategic purposes than their intervener counterparts (see Bayart 2010, Fisher 2014). Others, due to isolated, albeit spectacular, drug seizures that occurred in Mali in the late 2000s, insisted that "lessons learned" and "best practices" developed in response to the armed conflict in Colombia could be used in desert regions like Kidal and Timbuktu in northern Mali.

The tendency to conflate conflicts in the Global South is, of course, understandable. International bureaucrats, soldiers, peacekeepers, capacity-builders, development workers, diplomats, and various other members of what Lombard (2016) calls 'the good intentions crowd' were all drawing on their previous experiences as a prism to understand and respond to the complexity of the situation of war and armed conflict facing them. Such easily made comparisons have become widespread, regardless of how inaccurate, deeply problematic, and dangerous they can be. Tuareg rebels and jihadist armed groups were not the Taliban. The political economy of northern Mali has little in common with that of Afghanistan. The warfighting and levels of violence were nowhere near the same scale, nor was international involvement (see Malejacq and Sandor 2017). Yet, for these actors, there was always a way to see the relevance of making these comparisons. Some pointed to the security-development nexus, to counterinsurgency lessons, to the complexity of local dynamics and so on (Charbonneau 2019a). Afghanistan constituted a point of reference for understanding a conflict that they did not fully comprehend. Comparison became useful for international actors to justify intervention and to translate complexity into familiar solutions and programmes.

The practice of making conflict comparisons was not limited to international actors or news networks. It was also used by conflict actors to buttress their political positions in negotiations, to consolidate their control over fighters, and as calls for transnational rebel solidarity. For instance, in Mali, members of the *Coordination des Mouvements de l'Azawad* (CMA) argued that the implementation of the 2015 Bamako Peace and Reconciliation Accords must include amnesty clauses and formal integration and forgiveness processes for fighters currently aligned with armed Islamist groups based on models used in conflict settlements in Algeria, Mauritania, and Pakistan. As early as 2012, members of the secessionist *Mouvement National de Libération de l'Azawad* (MNLA) used the example of South Sudan or Eritrea as evidence of the nobility of their cause and the possibility of victory to indoctrinate their fighters, to gain support from populations living in areas they attempted to control, or to convince foreign researchers.

Apart from debates undertaken in the field of Comparative Historical Sociology in the 1990s on the politics of the comparative method and the effects that (mis)uses of history can induce to either support idiosyncratic or general sociological theories of change (see Tilly 1984, Kiser and Hechter 1991, Calhoun 1998), few studies of political violence or international interventions have examined the political effects of comparing civil wars, armed conflicts and conflict management (see Cramer 2007, pp. 49–138). Of course, several researchers compare armed conflicts and international interventions. Quantitative studies that draw on large datasets, amassing together constant variables from a diversity of contexts over time, are arguably dominant in the study of civil wars. Qualitative approaches in the study of political violence and international interventions also tend to mobilise comparisons in order to advance institutional and thematic knowledge of these issues, in the hope that they will be useful for policymakers. In this Special Issue, we argue that the act of comparing, however, primes for the production of categories, which, when mobilised, shape conflict management policies themselves, and thereby alter the trajectories of the very conflicts being compared. In other words, among other effects, comparisons have the power to produce and prioritise technical knowledge for the purposes of intervention. While there are important epistemological and methodological stakes associated with researchers engaging in political comparison, there are also important productive effects connected to epistemological and methodological avenues taken. Given the political dynamism of spaces of conflict and intervention, and that practitioners regularly seek out academic expertise, it is time that researchers coming from diverse research traditions to discuss the possibilities, limits, and the productive effects of our methodological and theoretical approaches to understanding and comparing armed conflict and intervention.

The everyday comparative method is a common, even banal, way of understanding and navigating the world. Everyone does it. As a scientific endeavour, the art of comparison (Lane 1997) seeks, like other methods, to explain and understand the world. Comparing is intimately linked to practices of categorisation since the concepts and typologies that it creates are necessary for the comparative exercise. And yet, it is precisely in these practices of categorisation that the comparative method encounters its possibilities, its limits, and its productive effects: in drawing lines and boundaries between the objects to be compared, comparison makes implicit or explicit judgements on the nature of these objects. As this Special Issue emphasises, this yields profound political consequences.

Performing Comparison: Performing the World

The power and power relations involved in establishing limits and boundaries to define, categorise, and understand the world through comparison are core themes of this Special Issue. In a way, the scholar is faced with the dilemma that Florian Kühn (2019, in this issue) identifies in the thought of Simone de Beauvoir: individuals try to make sense of the world and their place in it while knowing that any and all such attempts at understanding will shape and change both. This dilemma is not grounded in, or limited to, the scholarly predicament. The foundations and institutional mechanisms of the modern political order established, and continue to form, perform and impose conceptual and theoretical limits and boundaries. These foundations shape the very questions we ask, while setting aside others as impossibilities.

The state constitutes one such categorical boundary. Drawing on Mitchell's seminal work (1991), Jacob Mundy (2019, in this issue) identifies in his contribution on the production of the 'Middle East' as a political imaginary of violence how the political order established by the state produces distinctions between itself, society, and the international which generates power and resources that other forms of political order can hardly fathom. In comparing armed conflicts, in one way or another, the scholar faces the power of the state or, as Shapiro (2004, p. xi) put it, 'the "cognitive imperialism" of a state-centric social science'. In many ways, the state evinces a powerful grid of intelligibility that influences how we understand levels and units of analysis, thereby rendering possible the comparative endeavour, which in return contributes to the cognitive hegemony and reproduction of the state. As Mandy Turner (2019, in this issue) demonstrates in her article, comparing can indeed be about the state imposing, or trying to impose, its understanding of the world and its place or primacy in it – or as Bruno Charbonneau (2019b, in this issue) argues, to create a space for international intervention. Yet, it is also about the power of the state as a concept, as a unit of analysis, as an intellectual framework which have become so commonsensical that to call them into question, or to suggest that it is not so much 'how the world works' but an effect of power relations, is to transgress all types of boundaries. The state remains the starting point of armed conflict comparisons, and most attempts at going beyond the state always seem to bring us back to it (Walker 2010). Combined, the articles of this Special Issue certainly demonstrate this effect quite clearly.

State-centric assumptions manifest themselves in empirical investigations and methodological choices made. One example is through the use of conflict data-sets. In his article, Will Reno (2019, this issue) challenges the use of statistical methods relying on large data sets, which require a universal concept of the state in order to make comparisons to study civil wars and to support their claims to generalisable knowledge. Broad comparisons based on large-set data must ignore the context from which the data emerged. An immediate effect is the disregard for different logics of warfare. For example, Reno argues that the concept of state collapse – albeit contested – remains useful to examine how civil wars occur and evolve. State collapse produces 'a distinct context that shapes the uses of violence and the aims of important actors in these wars.' In other words, the context of collapsed states is distinct enough to challenge the validity of statistical-based methods. The possibilities of civil war comparisons depend largely on the *particular character* of *particular types of states* at (civil) war. The specificity of the collapsed state setting signifies that their dynamics are different from the civil wars that oppose the state to rebel groups.

Reno's analysis of the collapsed states in Mali, Somalia, and Iraq emphasises three common elements. First, strategies of regime protection undermine statebuilding efforts. To weaken the capacities of potential rivals for collective action, regimes empower non-state forces and subordinates by creating informal channels and networks separate from formal state institutions. The second element is a direct effect of the first: the systemic reliance on identity politics and what Reno calls 'neo-tribes' to consolidate and sustain the regime's grip on power. Third, 'neo-tribes' and their informal channels become connected to global networks once placed beyond the control of the regime. For Reno, such political relations remain unseen by large comparative data sets.

The benefits of the comparative method come first from asking the right questions and establishing a sound starting point: in Reno's case, the distinct logic of warfare of state collapse contexts. In her contribution, Róisín Read (2019, this issue) also emphasises the importance of context. She examines how comparative methods expunge key contextual knowledge when it comes to comparing sexual violence across conflicts or contexts. Unlike Reno, however, for Read the issue is not limited to methodology or the comparability of particular variables across time and space. Both quantitative and qualitative methods work in ways to efface the gender politics and relations inherent to sexual violence in conflict. Coding acts and events, as the building blocks of large data sets, require universal concepts to be organisable, measurable and ultimately useful to the bureaucratic mind. Yet, narrative and testimony-driven research also conjure up similar effects because it also requires a universal understanding of sexual violence in armed conflict in order to compare across countries and between cases. Moreover, the rise of evidence-based governance and thematic areas of international expertise are key to understanding how comparison can shape what we know about conflict-related sexual violence. By using categorised data, international public policy crowd comes to privilege technical over contextual knowledge because its expertise lends itself to transferability across countries, conflicts and contexts. In this way, 'international best practices', 'benchmarking' and other principles of the international policy trade translate into a cross-contextual coda that depersonalises histories and knowledges of armed conflict, sexual violence, interventions, and the like. Decision-making thus begins with a technocratic exercise that identifies what qualifies as valid knowledge in cases of war-time sexual violence as made possible by comparison and its power to obscure individualised contexts.

Read expertly illustrates and points to the politics of comparing conflict-related sexual violence that must be studied. Such an examination begins by interrogating definitions: what is sexual violence in armed conflict and who defines it? Defining the concept is unavoidably political. To define sexual violence is one, and to define it 'in conflict' is another. The latter implies that sexual violence can be carried out only by conflict actors, which excludes acts and structures of domestic violence, the exploitative acts of UN peacekeepers or humanitarians, and so on. Furthermore, it allows the claim that sexual violence is a weapon of war. All in all, the (political) need to define a universal understanding of conflict-related violence extracts it from gender relations and inequalities, while transforming it into an object of global governance.

Indeed, comparing can enable and enact the parameters of so-called 'appropriate' or 'tolerable' power relations. In her article, Mandy Turner (2019, this issue) dives right into the politics of comparative knowledge production. She is not so much concerned with where and how comparisons and their limits originate, but instead emphasises the need to analyse how metaphors and analogies are used politically, and how they sustain or challenge power relations. Turner argues that while the meaning-making effects of the comparative method are central aspects of our social reality, such effects do not constitute it. It is in analysing the truth-claims of metaphors and analogies that one can identify 'whether domination or emancipation is the desired object and outcome of their application'.

Turner's analysis is concerned with the political uses of analogies as a tactic of comparison in the context of the Israel-Palestine conflict. She focuses on two analogies. The first is the 'Holocaust-Hitler' analogy that Israel and its supporters deploy to portray the Israeli state 'as a beleaguered nation surrounded by anti-Semitic, Nazi sympathisers who see to destroy it as the Jewish homeland'. Strategically mobilising this metaphor serves a dual purpose: to unite the nation against the 'evil enemy', and to build international support and silence critics. The second is the 'apartheid analogy'. The predicament of the Palestinians is here compared to the victims of the racist, settler colonialism of the South African Apartheid regime. The comparison with South Africa is politically useful to make sense of the lived experiences of the Palestinian people. It also mobilises international support for the emancipation of Palestine and its people. Emphasising an analysis of who, and how, these analogies are enabled and enacted involves a research strategy that sheds light on the politics of comparing armed conflicts. In her article, conflict actors are the ones in the driver's seat when it comes to making comparisons between the Israel-Palestine conflict and two 'monstrous regimes'. The Israel-Palestine conflict context is a key case-study demonstrating how the practice of comparison is neither innocent nor without effects on conflict dynamics.

In turn, Bruno Charbonneau (2019b, this issue) builds on the critique of the state-centric comparative approach to demonstrate how comparing civil wars can graft itself the politics of international intervention. Conflict comparisons assume too much, or ignore too much, like the formation of prior political categories, imperial formations in particular. It is within the historical construction of a larger and largely ignored political context – called Francophone Africa – that comparisons are allowed between Francophone African states, and also between the Francophone African 'world' and other worlds. Charbonneau contends that the comparative method is an essential mode of knowing and being, but that its formal rules insisting on defining universal and static units of analysis cannot account for historical context. The proposed solution is 'a research strategy that emphasises the need to reveal and historicise the relations and processes that form or perform the very units of comparative analysis'. Here, Charbonneau argues against the methodological nationalism that plagues the comparative study of civil wars. Instead of taking the state as the given object of history and comparison, its historical formation must be placed at the centre of the analysis. This is meant to avoid the naturalisation of the state, and thus to reveal its colonial legacies, the power relations that constitute it, and the pervasive ways by which the postcolonial state is conceived as an object of international intervention.

Hence, the possibilities of comparing Côte d'Ivoire and Mali require the prior analysis of their construction as state units that constitute the 'world' of Francophone Africa. Charbonneau's analysis points to how 'the (French) privileged sphere of intervention is inextricably tied to the historical production of a privileged sphere of comparison called Francophone Africa which is inherited from French imperialism'. Partly defined as 'Francophone', the Ivorian and Malian states are recognised as members of a privileged sphere, as locations that can be distinguished from other places, and thus as historical formations that 'enable, enact and sustain a particular form of military intervention', more often than not led by France, to the exclusion of other geopolitical actors.

Jacob Mundy (2019, this issue) works along similar lines, sharing a common concern with Charbonneau about the production of a 'world' that serves particular interests and powers. In Mundy's case, he interrogates how and why the region called 'Middle East' is theorised as being prone to war, as emerging from and being reproduced by violent practices and processes. In fact, after showing that the region is no different than other regions, that it is no more and no less prone to war than other imagined geographies, he asks whether we need the idea of the Middle East at all. It is not clear what makes it a region, what its limits are, and what it is supposed to mean for thinking about war, conflict and intervention. For Mundy, there is a clear lack of theoretical engagement with the concept of region.

The answer lies elsewhere. He argues that regions are located at the nexus of the US academy and state power. Regions like the Middle East have served to strategize and to justify US hegemony. To ask, to suggest or to presume that the Middle East is more violent or prone to armed conflict is to assume that it has 'properties that are ontologically deeper and temporally prior to that violence'. The idea of the Middle East is racialised, and intertwined with assumptions about the so-called violent character of its people and social institutions. Put another way, to compare the Middle East to other regions is partly a boundary problem – where does it start and end? It is also about the processes and power relations that gave rise and reproduced the idea – an idea that has 'served to incorporate the Middle East within a global framework of political and economic relations underwriting and regenerating North Atlantic global supremacy'.

Romain Malejacq and Adam Sandor argue that the 'Sahelistan' discourse that conflates conflict dynamics in Afghanistan and Mali is widespread in Western media and policy circles. They show how such representations contribute to the adoption of one-size-fits-all conflict management policies. Drawing on extensive fieldwork in both places, they also demonstrate that these standard templates of intervention shape responses from local non-state armed actors, who manipulate foreign interveners for their own (violent) purposes in similar ways. Yet, they show that this convergence in armed group behaviour still manifests itself in important variations in widely different contexts, furthering strongmen autonomy in Afghanistan, while fostering armed group fragmentation in Mali.

Indeed, this Special Issue repeatedly demonstrates how power relations and comparative methods are intrinsically linked, raising the question whether comparison is possible in the first place. Florian Kühn (2019, this issue) argues that comparing conflicts will activate normative positions or assumptions that ascribe meaning to the conflicts under the scrutiny of comparative practice. For Kühn, comparing is meaning and world-making as it assigns order and place, creates hierarchies, and performs normative boundaries. As a practice of knowledge production, the comparative method is useful to support claims to expertise, to policy-relevant insights, and to guide action.

Kühn is particularly interested in those intervention actors who practice, use and benefit from making conflict comparisons. He identifies an 'intervention class' as an epistemic community that, through its involvement in conflict intervention, uses its members' experiences and perspectives to justify intervention, manage expectations, rewrite or erase history, categorise countries according to Eurocentric standards, and create their own authority as expert comparativists. Put another way, the intervention class does not merely create knowledge – no matter how accurate or biased – about the world, but establishes its own agency and its legitimacy as a conflict intervention actor.

Conclusion

The use of comparisons as a research method is valid, useful and at times powerful, but it is never innocent: comparing armed conflicts and interventions reproduces power relations and structures the very possibility of understanding the political world in either unquestioned or critical and transformational ways. Our Special Issue emphasises that there is always a prior or inherently political move before the comparison because of the necessity of having a beginning and of establishing the units of analysis under study. The problem of generalising armed conflicts obscures some context-specific conditions and dynamics when it highlights the similarities of others. In short, while comparing is closely linked to the theoretical and methodological possibilities of the comparative approach, it also constitutes a political move that orders and performs the social world. Hence, the practice of comparison has an impact on the production of scientific knowledge, on conflict dynamics themselves, and on the formulation and implementation of conflict management policy: comparison is inherently a practice of order-making.

Is comparison an exercise of domination and control? It can certainly be, as every article shows in its own way. Comparing to a 'normal' (Kühn), out of context (Reno), with quantitative methods (Read), or by putting a fence around a region (Mundy) are ways through which the comparative method can serve controlling and dominating agendas. Comparing can produce knowledge for and by the state (Charbonneau), which can banalise comparisons for purposes of instrumentalization by conflict actors (Turner). Comparing can sustain the will to increase the legibility and ability to discipline space and populations, the effacing of difference, the production of new, unintended actors and manifestations of self-fulfiling prophecy, and so on.

Does this mean that comparing armed conflicts is impossible or ill advised? Should we avoid it? Are our comparisons, in this issue, better than the 'bad' ones? These are questions that will be, and have been, raised, but they are traps. The issue is not about the possibility of comparison per se, because obviously no one will stop comparing armed conflicts. The issue is how comparisons make the transition into the world, how the abstract, conceptual, methodological, theoretical transform into practices or policies and how they affect power relations. Lots of things can get lost in comparing armed conflicts, and the canons of our disciplines often exacerbate the risks. Yet, as the contributions in this Special Issue argue, lots can also be learnt, especially if, instead of comparing armed conflicts per se, we emphasise comparing the processes, practices and relationships that constitute them as armed conflicts.

Acknowledgements

This Special Issue emerged from a workshop that aimed at unpacking the politics of comparing armed conflicts and their international responses. It was hosted by the Centre FrancoPaix in conflict resolution and peace missions, in Montreal on 13 October 2017. We want to thank all of our colleagues who contributed to the success of this special issue, whether in writing or at the Montreal workshop: Florian Kühn, Christian Leuprecht, Lotje de Vries, Jacob Mundy, Geneviève Parent, Róisín Read, Will Reno, Maxime Ricard, Mandy Turner and Marie-Joëlle Zahar. We also want to thank the staff of the Chaire Raoul-Dandurand at the Université du Québec à Montréal for organising the workshop. The workshop was also made possible by two grants, one from the Social Sciences and Humanities Research Council of Canada and another from the Canadian Defence Engagement Program. Adam Sandor thanks the Käte Hamburger Kolleg Centre for Global Co-operation Research at the University of Duisburg-Essen which provided a stellar intellectual space for academic production and financial support, without which this work would have struggled to complete.

Disclosure statement

No potential conflict of interest was reported by the authors.

Funding

This work was supported by the Canada's Defence Engagement Program; Social Sciences and Humanities Research Council of Canada.

ORCID

Bruno Charbonneau http://orcid.org/0000-0003-0373-7174

References

Bayart, J.F., 2010. Le piège de la lutte antiterroriste en Afrique de l'Ouest. *Sociétés Politiques Comparées*, 26. Available from: http://fasopo.org/sites/default/files/chro_n26.pdf.

BBC, 2013. UK to consider boosting French Mali operation support. *BBC News*, 22 Jan. Available from: http://www.bbc.co.uk/news/uk-politics-21136882 [Accessed 24 Nov 2019].

Calhoun, C., 1998. Explanation in historical sociology: narrative, general theory, and historically specific theory. *American Journal of Sociology*, 104 (3), 846–871.

Charbonneau, B., 2012. Le Mali n'est pas l'Afghanistan. *L'Actualité*. Available from: https://lactualite.com/politique/le-mali-nest-pas-lafghanistan-1/.

Charbonneau, B., 2019a. Intervention as counterinsurgency politics. *Conflict, Security and Development*, 19 (3), 309–314. doi:10.1080/14678802.2019.1608017

Charbonneau, B., 2019b. Privileged sphere of comparison: empire, methods and conflict intervention. *Civil Wars*, 21 (4).

Cramer, C., 2007. *Violence in developing countries: war, memory, progress*. Bloomington, IN: Indiana University Press.

Fisher, J., 2014. When it pays to be a 'fragile state': Uganda's use and abuse of a dubious concept. *Third World Quarterly*, 35 (2), 316–332. doi:10.1080/01436597.2014.878493

Kiser, E. and Hechter, M., 1991. The role of general theory in comparative-historical sociology. *American Journal of Sociology*, 97 (1), 1–30.

Kühn, F., 2019. Normative scaling and crisis knowledge: the problematic use of selective analogies to compare conflicts. *Civil Wars*, 21 (4).

Lane, R., 1997. *The art of comparative politics*. Boston: Allyn and Bacon.

Laurent, S., 2013. *Sahelistan*. Paris: Seuil.

Lombard, L., 2016. *State of rebellion: violence and intervention in the Central African Republic*. London: Zed Books.

Malejacq, R. and Sandor, A., 2017. Sahelistan? The risks and rewards of comparing the conflicts in Afghanistan and Mali. *In*: Paper presented at *The Politics of Comparing Armed Conflicts*, 13 October Montreal, Canada. Université du Québec à Montréal (UQÀM).

Michailof, S., 2015. *Africanistan: l'Afrique en crise va-t-elle se retrouver dans nos banlieues?* Paris: Fayard.

Mitchell, T., 1991. The limits of the state: beyond statist approaches and their critics. *American Political Science Review*, 85 (1), 77–96. doi:10.1017/S0003055400271451

Mundy, J., 2019. The Middle East is violence: on the limits of comparative approaches to the study of armed conflict. *Civil Wars*, 21 (4).

Read, R., 2019. Comparing conflict-related sexual violence: expertise, politics and documentation. *Civil Wars*, 21 (4).

Reno, W., 2019. The importance of context when comparing civil wars. *Civil Wars*, 21 (4).

Shapiro, M., 2004. *Methods and nations: cultural governance and the indigenous subject*. New York: Routledge.

Tilly, C., 1984. *Big comparisons, large processes, huge comparisons*. New York: Russell Sage.

Turner, M., 2019. Fanning the flames or a troubling truth? The politics of comparison in the Israel-Palestine conflict. *Civil Wars*, 21 (4).

Walker, R.B.J., 2010. *After the globe, before the world*. New York: Routledge.

The Importance of Context When Comparing Civil Wars

William Reno

ABSTRACT

Comparisons of civil wars based on attention to contextual categories do a better job of conceptualising key variables such as ethnic identity, resources, armed group membership, and concepts such as rebellion and negotiation than do studies based on large data sets and methodological individualism. This article shows how important variables and concepts apply in ways that are particular to conflicts that follow the collapse of centralised authority based on personalist networks. Comparisons with civil wars in highly bureaucratised states highlight these differences, and illustrate fallacies of assuming that variables and concepts transcend the broad historical sweep of civil wars.

This article ventures a comparative approach to uncover significant social relationships that shape civil wars.[1] The goal is to generate novel insights into conflicts that follow the collapse of personalist political systems that had maintained order behind the façade of formal bureaucratic administrations, or so-called 'collapsed states' (Patrick 2007). Though the concept of failed and collapsed states draws criticism for overgeneralisation (Call 2011), the argument here is that the concept of state collapse has a great deal of utility for comparing how and why civil wars occur and how they evolve. The process of state collapse creates a distinct context that shapes the uses of violence and the aims of important actors in these wars. Moreover, this context is distinct enough to challenge the validity of broad comparisons of civil wars that ignore the political contexts in which their main actors are situated.

Careful comparisons shed light on how pre-conflict processes of state collapse altered the social fabrics of communities to produce a distinct new social structure. Conflicts in Mali and elsewhere in the Sahel provide current examples of this transformation. Comparisons with other violent processes of state collapse reveal shared elements of these pre-conflict processes, albeit filtered through each locale's particular historical experience, geography, resource endowments, and evolving social norms. As a consequence, paying particular attention to the pre-conflict processes of state collapse uncovers shared factors and processes that otherwise might remain hidden. This approach generates important questions that otherwise might not be asked when a case is considered in solitude and on the basis of immediate events. It also enables one to generate observations across civil wars after the collapse of central state authority in ways that focus attention on the more significant elements of the granular details of particular cases.

For example, what is the real significance of 'tribes' and 'clans' in politics in collapsed states? These social-political groups play prominent roles in conflicts in collapsed states as diverse as Somalia, South Sudan, Afghanistan, Syria, Iraq, Yemen, Libya, as they do in Mali and elsewhere across the Sahel. These groups typically reflect local social connexions and their leaders often articulate deeply parochial political agendas. Though scholars usually do not intend to portray these groups as essential and unchanging, their treatment as a uniform variable in comparisons of large numbers of cases assumes that ethnic or clan groups are static identities of individuals and communities across long periods of time (Sambanis 2001, Fearon and Laitin 2003). Studies that focus on the processes of conflicts, such as how political entrepreneurs operate, the political uses of violence, and the survival strategies of those caught up in conflicts take a more dynamic view of these social categories as useful instruments for mobilising people for social, political and economic purposes, regardless of their prior origins (Snyder and Jervis 1999, Wood 2008). This approach is more accommodating to context and the possibility that particular categories of civil wars may have their own set of causes and processes that set them apart from other civil wars.

Social groups bearing ethnic and clan labels in many recent and ongoing conflicts simultaneously occupy key positions in global networks, such as smuggling and other illicit commerce, religious revivalist and extremist movements, and in diaspora politics. Rather than a reductionist notion of a revival of old solidarities, they are part of a historical process of identity formation and change as a consequence of strategies of pre-war governance and of wartime protagonists.[2] The situation in which heads of rebel groups that supposedly are rebelling against a government while simultaneously colluding in that government's principal channels of authority challenges a conventional view of the Tuareg rebellion in northern Mali as one case of a generalisable process in which political marginalisation from state power solidifies an ethnic group's common identity that then becomes an instrument that leaders use to rally the group to fight Wucherpfennig *et al.* 2012).

An alternative view rooted in the distinct character of state collapse depicts this rebellion not as direct contestation, but as the selective use of violence in an intra-elite political negotiation. Concepts such as 'state' and 'ethnicity' are variable in this approach. Separatist demands may actually be tactics in a struggle to control proceeds from illicit commerce such as drug trafficking and human smuggling that are integral to the exercise of political authority in which the 'state' consists of a bureaucratic façade that conceals personalist networks of formal and clandestine political and economic power (International Crisis Group 2018). Sudden and frequent shifts in alliances among a dizzying array of factions, and the seemingly friendly relations between leaders of opposing groups even while fighting, can seem puzzling to conventional approaches to the study of civil wars and call into question ethnic political solidarity and the meaning of 'separatist demands' in this context. Armed violence in this context is a tactic in the politics of intra-elite bargaining rather than a breakdown of politics and a resort to rebellion. This distinction is significant in addressing questions related to causes of conflicts, the significance of armed group fragmentation, and the roles of ethnicity, resources and so forth.

As products of distinctly contemporary processes of state and regime collapse, the argument below is that these categories should be analysed as part of a process of pre-conflict state collapse to better understand the roles they play during conflicts. This exercise in turn sheds light on processes that otherwise might be ignored, and points to specific actors who might be more important than they seem at first glance. As noted below, these social-political groups are more accurately conceptualised as 'neo-tribes' and 'neo-clans' rather than a revival of deeply rooted kinship organisations that predate the state.

This approach differs from studies that collect data across a wide array of constant variables to compare large numbers of conflicts across time. Many of these studies exhibit remarkable degrees of disagreement even on how to code basic concepts such as 'civil war,' and reify complex social processes like 'ethnicity' across time and place (Williams 2016, Baczko *et al.* 2017). That approach, for example, risks missing the fundamental difference between a concept like 'neo-tribe' that is a product of political strategies in collapsing states in which the exercise of political authority is deeply rooted in visible and clandestine interconnections beyond the collapsed state, and ethnonyms that may have been in use for over a century.

The analysis below starts with a focus on these common elements of social change that appear amidst the collapse of states. These common elements are the basis for an argument about the link between the governance strategies of regimes that intentionally promote the destruction of formal state structures and the distinct nature of conflicts that follow the collapse of these regimes. Attention then turns to a survey of several 21st century conflicts to illustrate the utility of applying these elements of a comparative approach to understand the underlying nature of these conflicts. That exercise shows how a comparative method illuminates key processes and actors. When applied to Mali and the greater Sahel, this exercise contributes to the foundation of worthwhile analysis, based as it is on knowing how to ask the right questions.

Common Features: The Social Structure of State Collapse

It is oft noted among scholars (if not always among policy makers and officials who press foreign governments to undertake reform) that the ruler who builds capable state institutions runs the risk that these institutions may serve as platforms for others to organise political opposition. defence against the latter often overshadows the pursuit of the former. Consequently, vulnerable rulers consciously undermine elements of the states that they are supposedly building (Huntington 1968, Bates 2008). This observation fits the concept of limited access orders that Douglass North *et al* understand as political systems that maintain their equilibriums in fundamentally different ways than contemporary institution-building orders. These are political systems in which states do not have secure monopolies of coercion. To control violence, their leaders instead divide up control over the economy to parcel out rents in return for support from elite factions (North *et al.* 2007). The ruler who fears armed challenges from these factions creates an armed force to protect against this threat. But the ruler may also have good reason to fear the ambitions of the members of that armed force that is supposed to provide protection (Feaver 1999). Coup-proofing to ensure regime survival in the short run crowds out state-building in the long run. These survival strategies involve parcelling out power to subordinates in personalist networks, usually distributed according to a logic that the architects and observers of these strategies alike interpret according to the metric of identity politics (Roessler 2011).

Rulers' fears for the security of their regimes and even their personal safety usually are well-founded. For example, across the 45 years from the 1956 independence of Sudan, the first in Sub-Saharan Africa's post World War II wave of independence, to 2001, 87.5 percent of Sub-Saharan African countries (42 of 48) experienced 'military interventions in politics', successful coups occurred in 62.5% (30 countries), and 37.5% (18 countries) experienced multiple coups (McGowan 2003). The Middle East across roughly the same period experienced successful coups in 9 of 16 countries (56%), with multiple coups in Syria (6), Iraq (4), and Algeria (3). Events since 2011 in the Middle East do little to make incumbent rulers there or in other regions feel particularly secure, and coups continue to succeed in Africa (10 from 2001 to 2015, including Mali in 2012), despite selective intervention on the part of regional organisations (most notably the Economic Community of West African States) to reverse some of them.[3]

Mid-twentieth century high modernist ideas of state construction were supposed to solve these problems. The massive extension of state services to citizens was intended to impose state authority over local strongmen, religious leaders and other local authorities (some officially recognised and others not) who government leaders saw as political competitors and obstacles to larger projects to reshape their societies and economies through industrialisation and urbanisation. The preference among rulers for one party rule usually involved national unity as a justification. Intellectual and activist movements (and quite a few governments) coalesced around socially inclusive radical ideologies, communism, pan-Africanism and pan-Arabism foremost among them, to promote even faster transformation at the expense of old categories of identity. Despite the popularity of these ideas, including in government circles, this reshaping of state relations with citizens ultimately generated what many rulers regarded as subversive threats as citizens began to mobilise in ways that alarmed regimes (Bueno de Mesquita and Smith 2012). Extraordinary oil revenues and foreign aid windfalls gave some rulers the capacity to double-down on an institutional-based expansion of state intervention into society, though usually in ways that left citizens deeply dependent upon the state. This dependence, whether in the context of institution-building or in the increasingly common institution-suppressing regime strategies, came to play a significant role when state rulers shifted the terms of their engagements with their societies.

The first common element in strategies of regime protection is the empowerment of subordinates through channels separate from the formal institutions of the state. Put differently, this strategy leads rulers to burn down the institutions of their own states, particularly among rulers in the regions noted above for the mid-century onset of serious risks of coups d'états and other violent domestic threats to incumbent regimes. This shift away from classic state-building efforts addressed the threat that competent managers could hijack increasingly effective (and therefore potentially autonomous) bureaucracies to pursue their own political interests. From the perspective of many regimes in Africa and the Middle East it seemed safer to allow supporters to exploit ostensibly private and even illicit commercial opportunities that are dependent upon the discretion of the ruler than it is to promote the expansion of an efficient and politically autonomous commercial class.[4] The integration into these personal networks of otherwise purely parochial issues such as local land tenure disputes, and even family and personal rivalries, attaches yet more dimensions of everyday life to the ruler's discretion. From the perspective of many regimes, this is a radical yet necessary departure from the strategies for exercising authority typically employed in the 1960s and into the 1980s.

Extraversion is a central feature of this transition, imparting a global dimension to these alternative channels for the exercise of authority (Bayart 1999). Extraversion describes how rulers and their associates actively participate in processes that outwardly appear to be central to the creation and maintenance of a country's dependence upon and subordination to foreign political and economic actors. But the dynamics of this dependence rather than the outward appearance of subordination are more important for analysing this situation in the context of regime strategies.

This approach draws attention to a historical reading of participation and movement in this set of global connexions. Seen in this light, extraversion consists of a set of rent-seeking strategies, such as using the selective enforcement of laws to shelter international criminals and share in their business profits, sell diplomatic support, divert foreign aid, participate in foreign-financed peacekeeping operations to collect and distribute resources to political supporters in the army and so forth. The willingness of foreigners to recognise the sovereignty of these states is a defining feature of the ruler's ability to manage external dependence through rents, which in turn are used to manage the dependence of domestic subordinates (Bayart 1989). But once regime control collapses, the material resources and the various foreign connexions associated with extraversion are turned to the pursuits of the goals of the armed groups that arise out of this context.

The second common element in strategies of regime protection is the systematic regime reliance on 'neo-tribal' (and in some instances, sectarian) identity as a principal framework for ordering politics. The prefix 'neo' better reflects the reality of the relationship of these social categories to state authority. There is a great deal of social distance between identity categories in the distant past and how regimes use them as part of a strategy of shifting the exercise of authority away from formal institutions to networks of personal patronage. Neo-tribes and neo-clans lack many of the social practices and ethical norms that most observers subscribed to tribes and clan systems that were undergoing extensive atrophy from the middle of the twentieth century's modernist phase of state-building (King 2001). At that time, political leaders commonly identified such categories as obstacles to modernity and national unity, and sought to undermine their influence.

It is common across regimes that purposely collude in the collapse of their own states' institutions to elevate supposed tribal and religious leaders to prominence. This dramatic development, often after a particularly acute threat to a regime, such as an uprising or other major security threat, can look like a rebirth of suppressed or dormant traditions. Turns of this sort took place in Syria in the early 1980s with the bloody suppression of an Islamist movement, in Libya in the early 1970s after a series of coup attempts, and in many African states after the demonstration effects of a dramatic increase in coups d'états in the mid-1960s. This approach mirrors the longstanding counterinsurgency tactic of promoting the formation of tribal militias to provide intelligence and as force multipliers for regular armed forces (Porch 2013). Yet this is hardly a triumph of tradition. These new 'traditional' leaders arise out of a distinctly contemporary context. Their prominence is a consequence of a regime's deliberate political strategy to exploit and manipulate social divisions, couched in familiar categories and norms that are associated with the past. These categories usually owe more to the enormous stresses that attend the abandonment of the modernist state-building process and a ruler's selective incorporation of communities into personalist networks as to an imagined distant past.

These categories thus in no way are revivals and rediscoveries of latent authentic kinship. They typically include notably irreligious and non-tribal individuals and communities that are then transformed through this political process. Another critical feature is the connexion of these social groups to global networks, partly as a consequence of a ruler's first strategy of shifting the exercise of authority from state institutions to these alternative channels. These networks of commercial exchange (including smuggling), religious education and so forth remain dependent upon a ruler's discretion, so long as a ruler is able to punish insubordination and control who speaks for the neo-tribe or neo-clan.

The third common element of regime protection strategies is the seemingly paradoxical connexion between 'tribes' and the global networks of dependence that are at the centre of the collapsed state ruler's strategy of extraversion. A prominent manifestation of this paradoxical connexion appears in the observation that all jihadist 'emirates' are located in 'tribal' lands. While not all jihadists come from these neo-tribes, all jihadists who actually control territory play the politics of 'tribe' as a central element of their own strategies of authority and control (Collombier and Roy 2018).

Those who rule collapsing states do not set out to create jihadist enclaves. These rulers are key actors in this process, however, as they lay the foundations of this juncture of parochial and global action in very concrete and socially defined ways. A series of decentralised and globally connected personalist networks work as instruments of regime control so long as a ruler is able to supervise their subordinates who are able to operate across the external and internal dimensions of these networks. But sudden shocks, such as an international armed intervention, a serious economic crisis, or foreign political pressure to hold genuinely competitive elections, undermine centralised control over these networks. Then key members of these networks gain much greater autonomy to set their own terms in their relations with community members and global networks.

Neo-traditional leaders typically exploit the opportunities associated with regime weakness and collapse to promote their personal political fortunes. They often push aside or kill off community leaders who previously played conventional intermediary roles, such as helping community members get state services or protecting communities from overbearing state institutions. Regime weakness and demise unleash enterprising neo-traditional leaders to appropriate their connexions to global networks to serve their own political ambitions. This reconfiguration of neo-traditional parochialism and global networks reverses the old regime-dominated hierarchy. The exploitation of global connexions in this context of state *and* regime collapse amounts to a 'repositioning in the regional geostrategic context allowing the tribe to contest the state,' such as it is, 'no longer from below but from above' (Roy 2018, p. 7)

The broad parameters of conflict in the Sahel and in other parts of Africa and in the Middle East can be seen as a product of a crisis of modernist state construction rather than a reversion to a pre-state past. These contexts also exhibit common patterns in terms of the relationship of political actors on the ground and global insurgents. Comparisons among these conflicts thus shed light on how local socio-political organisations that appear to be intensely parochial in outlook and interests become attached (or reject) global insurgents.

Pointing towards the Right Questions: The Benefit of Comparative Analysis

Corruption is a problem in many countries, but a ruler's intentional subversion of the state is distinct from simply using bureaucracies as vehicles to distribute patronage or personal enrichment. Indeed, many leaders who presided over the collapse of their own state's institutions later were notable in earlier times for their commitments to state-building projects.

Burning down the State

Somalia's President Siyad Barre, installed in a coup d'état on 21 October 1969, stated out as a state-builder. He announced a 'Scientific Socialist' programme, forbade official references to clan identity, and promised to unify all Somali language speakers in a single state, which was to include ethnic Somalis in neighbouring states (Markakis 1987). In 1977, Somalia's army invaded Ethiopia to annexe these ethnic kinsmen, a rare post World War Two example of a classic state-building project through warfare. By March 1978, Ethiopian counter-attacks reached Somali territory, and in April disgruntled army officers attempted a coup d'état.

Barre's response was immediate and targeted the institutions of his own state. He started with purges and reprisal killings aimed at members of the coup plotters' communities (Kapteijms 2013). Over the longer-term, Barre relied heavily on a strategy of extraversion to pay for an expanding political network as he let state institutions whither. Foreign aid totalled $2.8 billion between 1977 and 1989, making Somali people Africa's greatest per capita beneficiaries of aid at that time. But social service provision was low on Barre's agenda, and accounted for less than one percent of budgeted spending in 1989 (World Bank 1991). Instead, these resources, along with economic reforms, helped the migration of regime favourites into commerce and gave them land in politically suspect communities as part of a US-supported drive to promote commercial agriculture (Sommer 1994). The extreme dependence of the country on foreign aid for 90 percent of official development spending reinforced Barre's central role in all avenues of economic life – as manager of formal state allocations and as patron to political allies – at the expense of autonomous entrepreneurial efforts. These resources contributed to what Alex De Waal (2015) calls the 'political budget' in a monetized system of patronage in which the ruler trades political loyalty for opportunities to exploit economic opportunities in a global patronage order. Barre's domination of these economic channels entrenched his personal power in communities that received foreign aid, imposing his political network over formal institutional procedures, illicit and informal commercial networks, and incumbent community authorities.[5]

By the time the regime collapsed in January 1991, Barre had burned down his state with help from foreign aid that he used into building his personal political network.[6] But the Cold War ended and the US Congress questioned Barre's commitment to human rights – his regime killed an estimated 50,000 to 60,000 civilians in the city of Hargeisa – so aid was cut off (Africa Watch 1990). With the sudden loss of external resources, Barre lost control over subordinates in his political network, which led to the collapse of his regime in 1991. These networks survived as members appropriated these connexions for their own benefit. This development shows the value of examining the social process of state collapse as a distinct generalisable phenomenon to understand the social configuration of subsequent conflict.

Iraq president Saddam Hussein traversed a similar path. He began in the 1970s as a state-builder with help from oil revenues that increased from $575 million in 1972 to $26.5 billion in 1980 (Sluglert and Farouk-Sluglert 1990). Government policies and oil wealth brought a massive expansion of state services and transformative infrastructure construction projects. A huge army and massive security services employed half a million people and played an important role in Saddam's project of using state and Ba'ath party institutions to dominate Iraq's people (Pollack 2002). Land reforms in the 1970s targeted the position of sheikhs, and ruling Ba'ath party cadres were sent to the countryside to help eradicate backward norms in an explicitly anti-traditional campaign (Baram 1997, p. 3).

Saddam's strategy by the 1990s had undergone a significant shift. Iraq's invasion of Iran in 1980 led to eight years of war and 600,000 Iraqi casualties and the ill-fated invasion of Kuwait in August 1990 and rout at the hands of the US-led coalition weakened the regime. A more serious threat, however, came in 1991 from the simultaneous uprisings in the Shia south and Kurdish north, followed by broad UN sanctions in the 1990s that upended the old strategy of using oil revenues to extend state institutional engagement and control of the population.

Backers of the UN sanctions expected Saddam to fall from power at the hands of domestic opponents. Instead, he shifted the basis of his regime's power from the formal institutions of the state to manipulation of the effects of sanctions. Rationing and subsidies became instruments to reward supporters, punish perceived opponents, and make everyone else dependent upon regime favour. Saddam exploited the growing informal economy to permit political insiders to exploit illicit economic opportunities while prosecuting those who acted on their own. Civil service salaries shrank to the point of insignificance, forcing many of these people to join others in seeking places in networks in the illicit and informal economies that had political protection from regime insiders, or to turn their government offices into 'profit centres' under a government policy that effectively legalised bribes (Rigaud 2001). Security services, now supervised by the president's son, continued to employ large numbers of people, while new security agencies were created to monitor the activities of the others.

As in Somalia, these regime networks beyond the formal state were instrumental in shaping patterns of the conflict that followed the regime's collapse. After 2003, these networks were central to the insurgency against the US-led coalition and Iraq's new government (Munson 2009). As in cases of regime-led state collapse elsewhere, the resulting social structure held many important clues as to who and what came next; a vital social architecture that should be of great interest to scholars and practitioners.

Applied to Mali, this approach starts with attention to regime efforts after independence in 1960 to build what they conceived as a modern bureaucratic state. Officials employed considerable symbolic resources – the legacies of Timbuktu, Gao, Djenné – and the historic Mali Empire of Sundiata Keita (1235–1255) from which the country took its name. Mali's first president, Modibo Keita (1960–1968), claimed direct lineage of the imperial family, combining this claim to state-building legitimacy with a socialist orientation. Officials held that economic development and state-building would succeed 'only if the mentality of the rural population could be transformed from a backward traditional outlook on production and society to a modern rational one' (Lecocq 2010, p. 310–311). But the rapid expansion of the state and efforts to kerb the power of local authorities rapidly led to bankruptcy and resistance, and then to a coup in 1968.

Keita's successors, including General Amadou Toumani Traoré (ATT), leader of a coup d'état in March 1991 and a subsequent democratic transition, ruled in a very different fashion. For example, Mali's government contributed virtually no funds for primary health care, leaving foreign donors to fund the entirely of that service (Peters *et al.* 1999). Public education underwent rapid collapse from the mid-1980s, replaced with Koranic and other private institutions (Gérard 1992). This decline continued through the 1990s, despite increased donor funding for education. Foreign pressure to decentralise state administration actually provided further opportunities to shift the basis of authority away from dying formal state institutions and into personal networks. This policy equipped the regime to target local 'Big Men' to co-opt them into a patronage system that gave them access to the protection and rent-seeking opportunities of the central state and license to engage in illicit activities in return for political support (Bøås 2012). Historically these networks in the north were closely linked with smuggling, while some of them by the mid-2000s were deeply implicated in drug trafficking and kidnapping.[7]

These networks, central elements of the dissolution of formal state institutions, were integral to the regime's exercise of authority. In Mali, as elsewhere, these networks eventually challenged the ruler's hierarchy. The value of this comparative method lies in the guidance that this approach provides in examining the social structures of these conflicts. In particular, it draws the analyst's attention to elements of personalist rule such as 'illicit' networks (as seen from the perspective from outside these systems of authority) and the strategies of extraversion that shape the character of warfare that follows state collapse. While the association of natural resources and warfare may remind readers of analyses of Paul Collier, Jeremy Weinstein and others who identify resources as primary causes and shapers of conflict (Collier 2000, Weinstein 2007), the analysis here privileges the social structure of state collapse and the role that it plays in shaping how resources are used in warfare once a ruler's control over personalist networks fragments.

Despite local contingencies and circumstances, these social structures of collapsed states share several key features that are important to the analyst: (1), the networks of opposition are integral to the networks of the political establishment, (2), members of these networks hedge their bets and often exercise influence within the political establishment at the same time that they oppose it, and (3), foreign efforts to shore up the state (in a conventional form of this concept) often contributes to the empowerment of these networks.

Neo-tradition

Turning again to Iraq, Saddam's promotion of tribe and Islam from the late 1980s was not a revival of dormant or innate forces. It was a distinctly modern phenomenon, a consequence of regime efforts to exploit labels from the past as part of a strategy to empower politically subservient bosses at the expense of old local establishments. Newly appointed 'tribal' leaders were selected with an eye towards exacerbating local tensions to atomise community capacities for collective action. Neo-tribes filled this void, extending networks of dependence as remaining state services dwindled and institutions crumbled (Jabar 2003).

Saddam's efforts to pose as a patron of Islam in the late 1980s and particularly after the revolts of 1991 were aimed at undermining a religious revivalist movement centred on Najaf and establishing new networks that the regime could use to collect information (Helpont 2018). As low-level neo-tribal leaders helped the regime to suppress revolts (once these individuals perceived the regime to be winning), regime-appointed 'sheiks' provided additional conduits of information (Al-Khafaji 1994). Many of these neo-religious and neo-tribal leaders took advantage of illicit opportunities to build their own powerbases. The regime cultivated this transformation, provided these individuals focused on competition with one another and relied upon the regime to maintain their local positions. These neighbourhood-level tensions were a valuable resource for the regime, which could become the arbiter in intensified inter-communal jockeying around otherwise purely local and parochial issues. The regime's strategy of extraversion also linked these individuals to networks that led to the creation of institutions such as 'criminal mosques' that ran mafia-style protection rackets and illicit business operations. These mosque leaders and neo-tribal bosses operated much as neighbourhood warlords, many of whom later readily connected to outreach efforts from Iran and various Salafist groups as well as foreign criminal gangs once the US-led invasion toppled Saddam's regime in 2003.

The generalisability of this model is reflected in the trajectory of Somali politics to the 1991 collapse of President Siad Barre's government. Installed in office as a result of a coup d'état in 1969, Barre immediately outlined a crash literacy program. In 1970 he announced his programme of 'Scientific Socialism' that included massive state-led development projects and banned references to one's clan and ethnic origins in official documents and public statements. Between 1970 and 1975, government revenues doubled and social services expanded. Barre also built one of Africa's largest armies, and in 1977, invaded Ethiopia in an attempt to annexe areas inhabited by ethnic Somalis. A year later, his army was in retreat, and a series of coup attempts followed. Barre's defensive strategy included effectively dismantling his army into a collection of militias under the control of regime insiders who were selected on the basis of their clan identities. As noted above, Barre maintained their political loyalties through regulating their access to illicit economic opportunities and selective exemption from prosecution. (Miller 1981, Jamal 1988). Meanwhile, by1989, civil servants were paid only three to four percent of the real value of their 1975 salaries, and had to support themselves through corruption, bribes and kickbacks (Gregory 1990, p. 12).

Like Saddam, Barre had effectively burned down his own state after he discovered that modernising government administration and mobilising citizens for economic development could be threatening to his own rule.

The state itself became the regime's target, as the ruler built a system of personalist rule rooted in illicit economies behind the façade of formal statehood. It is these neo-traditional networks of family ties, ethnic militias and illicit economic enterprises behind the collapse of the pre-war state that shape the patterns and character of conflict. This strategy goes beyond conventional 'divide-and-rule', as it also involves decentralising the exercise of violence and access to economic opportunities. This works so long as the ruler can exercise personal control over this system. Once that central control collapses, the intended underlying fragmentation comes to the fore. International actors can play a role in solidifying this neo-tribalism through selectively empowering identity-based armed groups in ceasefires and power-sharing agreements (Tull and Mehler 2005). Thus a sensible comparative method that is suited to this distinct category of conflict that follows state collapse focuses on the identification and analysis of this social structure, and the logic of actors' behaviour that, the legacy of this process of failed modernisation and state collapse.

In Mali, foreign donor and creditor-imposed administrative decentralisation in the mid-2000s and insistence that foreign aid be used to support locally-run administrations provided the shock that undermined the personalist regime that earlier hollowed out formal state institutions. The country's rulers had ruled with help from a network of armed ethnic and clan militias that they used to exploit inter-group rivalries. They exacerbated these rivalries through permitting selective access to illicit commercial networks (Briscoe 2014). But competitive elections and decentralisation, coupled with their existing illicit economic activities, gave local strongmen the capacity to recruit and provide for their supporters independently on the president's wishes (Scheele 2009). This shift fragmented the ruler's personalist network and contributed to the 2012 coup that led to a civil war. As in Iraq and Somalia, the neo-tribal 'rebel' leaders that appeared in the conflict from 2012 really were rooted in the pre-conflict political system's networks.

Therefore, in comparative terms it is more appropriate to conceptualise these actors and their categories as reflective of networks of personalist rule and the collapse of a state (or more accurately, the personalist authority that had replaced formal state institutions), rather than deep-rooted ethnic or clan tensions or a battle to control resources in the first instance. The framework of the collapsed state category also helps to make sense of neo-tribal leaders who seem to have remarkably fluid loyalties to the extent of suddenly collaborating with members of the government and other groups that they are fighting to exploit commercial opportunities within these networks.[8] If this political field of contention is distinctly unstable (though within a clear and predictable logic of action) within the collapsed state context, comparisons with other categories of civil war conflict may help to explain why it is so hard to reach conclusive negotiated settlements. Likewise, the sensible negotiator would do well to take into account these distinctive actors and relationships in an effort to reach a realistic settlement.

Neo-tribes and Global Networks

These examples illustrate the mechanisms through which the social organisations of state failure such as neo-tribes connect to global networks through their appropriation of the regime's strategies of extraversion. These global connexions also bring contact with other networks, such as Salafi and other mobile preachers. Somalia and the Afghan-Pakistan border regions provide a good comparative illustration of how these newer religious networks merge with the existing smuggling and other commercial networks. Members of these existing networks discovered that joining Salafi global networks benefitted their business operations. This membership introduced them to new overseas contacts they could trust and gave them an organisational mechanism to conduct business and enforce contracts within their own communities (Ahmad 2017).

In general comparisons within the category of failed states, developments of this sort again draw attention to the importance of context; It is not Diasporas and international religious networks that shape the character of conflict as it is the extraverted networks of state collapse – the neo-clans and neo-tribes – that then are particularly well-suited to connect with transnational actors that are able to adapt to this fragmented political environment.

This mechanism through which neo-tribal and other neo-traditional social organisations connect to global networks sheds light on how and why groups like al-Qaeda and Islamic States develop links with these social organisations. The importance of this relationship is borne out in the dominance of these connexions beyond the Sahel, such as in Libya, Yemen, Syria and Iraq. All of these global networks have a 'tribal' policy in which the outside actor and the neo-tribal actor are able to identify a shared interest. Since most neo-tribe leaders arrived in their positions as consequences of the state regime's post-institutional social strategy, they tend to come from outside of old established community circles. As upstarts, they need the help of this external connexion to graft onto the networks that they inherited from their places in the old regime's strategy.

connexions to new global networks give these upstarts their own access to resources and prestige independent of the old local elite. Such individuals find opportunity amidst chaos, since they are able to use these new external links to compensate for disreputable backgrounds, poor education, and limited experience. A good example of one such individual is Moqtada al-Sadr in Iraq, an upstart who was able to benefit from Iranian technical assistance and political support after 2003 to jump into an existing community factional rift, prevail, and construct the Mehdi Army militia (Krohley

2015). Mali and Somalia have plenty of their own examples of these political entrepreneurs who build militias along these politicised identity lines that then map onto the pre-conflict regime's strategies of control.

Better Comparisons Lead to Better Questions

This approach to comparisons of civil wars in collapsed states addresses the broader comparative enterprise as it is conventionally applied to the study of civil wars. If state collapse produces a distinct logic of warfare – centred on the pre-war regime's networks of authority, for example – is a post-collapse a 'civil war' in the same sense as, say, the conflict between Sri Lanka's Tamil Tigers and the Sri Lankan government up to the Tigers' decisive military defeat in 2009? The social context of Sri Lanka's civil war was far removed from the personalist politics and social impacts of state collapse, and included a dense hierarchy of bureaucratic state agencies that became the basis for the Tigers' rebel governance structures in places that they controlled and from which they drew revenues (Mampilly 2015). This practice made these rebels vulnerable to a Sri Lankan government offensive that targeted rebel bureaucracies and the rebels' own hierarchically organised armed force. The underpinnings of the Sri Lankan separatist conflict and its' duality of state vs rebel violence is absent in Congo's ongoing conflict that began in 1996. The rebels, local militias, elements of a national army, and freelancing bureaucrats with their own powerbases engaged in this conflict are legacies of the pre-war political context of a collapsed state that President Mobutu's personalist networks of authority had largely superseded (Sweet 2018). The collection and uses of resources play very different roles in rebel strategies in Congo's collapsed state when compared to a civil war in which a rebel group fashions itself as a highly bureaucratised state-in-formation to fight a bureaucratic incumbent government.

The broader importance of context for comparisons of civil wars lies in what context means when viewed across the historical sweep of civil wars. Keepers of large data sets, such as the Correlates of War project (cataloguing wars from 1816 to the present) have expanded their typologies of civil wars to include such categories as 'non-state' and 'intercommunal' wars (Sarkees 2010). But even within a category like 'non-state' wars, conflicts such as War #1500: The First Maori Tribal War of 1818–1824 involves a radically different context than #1594: The Hemda - Lendu War of 1999-2005 (in Congo) (Sarkees 2010). Variables such as 'ethnicity,' 'resources' and the tactics and strategies of armed groups operate in radically different ways across these different authority structures that shaped fighting, external links, economic relationships, and so forth. In Congo's war, for example, the pre-war regime manipulated communities' access to land and citizenship rights in ways that were designed to fragment identity groups at the same time as intensifying their roles in politics. This was a deliberate strategy to draw local conflicts into a regime-managed struggle for economic and political resources to prevent a collective challenge to the regime (Vlassenroot and Raeymaekers 2004). Surely this process of state collapse set in motion a very different process of identity formation and mobilisation than the early 19th century wars at the start of massive alien conquest and occupation in what as to become New Zealand. Contemporary state collapse also sets in motion a distinct category of internationalised civil war that complicates efforts to make uniform comparisons across historical eras.

These differences at a conceptual level challenges the assumption that discreet variables that are associated with civil wars can be meaningfully standardised across big data sets that span a wide array of contexts. These differences also show that civil wars are reflective of the character of the states that rebels challenge. While not necessarily deterministic (such as in the highly centralised Islamic State's challenge to the highly politically fragmented government of Iraq and the dozens of militias that make up its' armed forces), there does appear to be a correlation between the organisational character of a state's authority structures and the character of the groups that fight in a civil war. For example, the most organisationally hierarchical rebel groups such as the Tigrayan People's Liberation Front (Ethiopia) and the Rwandan Patriotic Front fought and overthrew governments in these African states that have among the most extensive historical experiences with conventional bureaucratic state structures that reach deep into their pre-colonial pasts. By contrast, civil wars in collapsed states like Liberia, Somalia and Congo occurred in contexts of deep historical experiences of governance organised around highly extraverted patron-client networks. The plethora of rebel groups, their fragmentation and frequent shifts in alliances and tactical uses of violence in intra-elite negotiations are reflections of the strategies of authority in the personalist regimes before the ruler's central command of these systems was broken.

Moving context to the foreground in comparisons of civil wars in collapsed states lead to questions that are central to how politics works in collapsed state contexts that otherwise might be missed in broad comparisons across categories of civil wars. Some of these questions include:

- How did the process of state collapse and regime strategies shape a local social structure?
- How does this social structure influence for what and how people fight?
- How did extraversion connect elements of this social structure to global networks?
- How and why did leaders of neo-tribe social organisations emerge (i.e., what connexions and skills enabled their emergence)?

This approach lays a foundation for a comparative approach to the study of civil wars because it helps to identify local and global factors that otherwise remain hidden in comparisons based upon large comparative data sets that simplify dynamic processes, such as ethnic identity, resource flows, and the basis for individual and group strategies. As comparisons within a category such as civil wars in the context of state collapse, and comparisons between this and other contextual categories show, these factors and processes can play radically different roles at this level of analysis of civil wars. This consideration of context draws attention to specific actors, their political strategies, and connexions to important local and global networks that might otherwise be opaque or hidden in more conventional analyses and frameworks for comparison. These contextual factors are important to know, not just from an academic perspective. Policy-makers who intervene in these civil wars also benefit from context-specific knowledge that can help to identify key actors and processes to at least have a more realistic chance of influencing them towards achieving the aims of conflict reduction.

Notes

1. This essay is an outcome of the Research Council of Norway project 250173, 'Political Economy of Jihad.'
2. For an influential critique of the reductionist uses of indices of ethnic and social fragmentation when applied to armed conflict, see Cramer (2002).
3. Author's calculation.
4. Collier (2007) explains from an economist's perspective the incentives for rulers to rule poorly.
5. Somali Ministry of National Planning (1987, p. 10). Foreign consultants and advisors often wrote ministry reports evaluating project development, including critical ones like this, as formal state capacity shrank.
6. Among the best analyses of this regime is Compagnon (1995).
7. On smuggling networks, Scheele (2012); for the expansion of illicit activities, Lacher (2012).
8. As in Mali: (Shurkin *et al.* 2017, p. 54–56).

Disclosure statement

No potential conflict of interest was reported by the author.

Funding

This work was supported by the Norges Forskningsråd [250173].

ORCID

William Reno http://orcid.org/0000-0002-7592-9610

References

Africa Watch, 1990. *Somalia: a government at war with its own people*. New York: Africa Watch.

Ahmad, A., 2017. *Jihad & co.: black markets and islamist power*. New York: Oxford University Press.

Al-Khafaji, I., 1994. State terror and the degradation of politics. *In*: F. Hazelton, ed. *Iraq since the end of the Gulf War: prospects for democracy*. London: Zed, 20–31.

Baczko, A., Dorronsoro, G., and Quesnay, A., 2017. *Civil war in Syria: mobilization and competing social orders*. New York: Cambridge University Press.

Baram, A., 1997. Neo-tribalism in Iraq: Saddam Hussein's tribal policies, 1991–1996. *International Journal of Middle East Studies*, 29 (1), 1–31. doi:10.1017/S0020743800064138

Bates, R., 2008. *When things fell apart: state failure in late-century Africa*. New York: Cambridge University Press.

Bayart, J.-F., 1989. *L'État en Afrique*. Paris: Fayard.

Bayart, J.-F., 1999. L'Afrique dans le monde: une histoire d'extraversion. *Critique Internationale*, 4 (5), 97–120. doi:10.3917/crii.p1999.5n1.0097

Bøås, M., 2012. Castles in the sand: informal networks and power brokers in the northern Mali periphery. *In*: M. Utas, ed. *African conflicts and informal power: big men and networks*. London: Zed Books, 119–134.

Briscoe, I., 2014. *Crime after jihad: armed groups, the state and illicit business in post-conflict Mali*. The Hague: Clingendael Institute.

Call, C., 2011. Beyond the 'failed state': toward conceptual alternatives. *European Journal of International Relations*, 17 (2), 303–326. doi:10.1177/1354066109353137

Collier, P., 2000. Rebellion as a quasi-criminal activity. *Journal of Conflict Resolution*, 44 (6), 839–853. doi:10.1177/0022002700044006008

Collier, P., 2007. *War, guns and votes: democracy in dangerous places*. New York: HarperCollins.

Collombier, V. and Roy, O., eds., 2018. *Tribes and global jihadism*. London: Hurst & Company.

Compagnon, D., 1995. Ressources politiques, régulation autoritaire et domination personnelle: le régime Siyyad Barre en Somalie (1969–1991). Thèse intitulée Université de Pau et des Pays de l'Adour.

Cramer, C., 2002. *Homo economicus* goes to war: methodological individualism, rational choice, and the political economy of war. *World Development*, 30 (11), 1845–1864. doi:10.1016/S0305-750X(02)00120-1

de Mesquita, B.B. and Smith, A., 2012. *The dictator's handbook: why bad behavior is almost always good politics*. New York: PublicAffairs.

De Waal, A., 2015. *The real politics of the Horn of Africa*. Cambridge: Polity Press.

Fearon, J. and Laitin, D., 2003. Ethnicity, insurgency, and civil war. *American Political Science Review*, 97 (1), 75–90. doi:10.1017/50003055403000534

Feaver, P., 1999. Civil-military relations. *Annual Review of Political Science*, 2, 211–241. doi:10.1146/annurev.polisci.2.1.211

Gérard, E., 1992. Entre état et population: l'école et l'éducation en devenir. *Politique Africaine*, 47, 59–69. Available from: http://www.documentation.ird.fr/hor/fdi:36895

Gregory, P., 1990. *Somali civil service reform*. Mogadishu: USAID, mimeo.

Helpont, S., 2018. *Compulsion in religion: Saddam Hussein, Islam, and the roots of insurgency in Iraq*. New York: Oxford University Press.

Huntington, S., 1968. *Political order in changing societies*. New Haven: Yale University Press.

International Crisis Group, 2018. *Narcotrafic, violence et politique au Nord du Mali*. Brussels: International Crisis Group.

Jabar, F., 2003. Sheikhs and ideologues: deconstruction and reconstruction of tribes under patrimonial totalitarianism in Iraq, 1968–1998. *In*: F. Jabar and H. Dawod, eds. *Tribes and power: nationalism and ethnicity in the Middle East*. London: Saqi, 69–79.

Jamal, V., 1988. Somalia: understanding an unconventional economy. *Development & Change*, 19 (2), 203–265. doi:10.1111/j.1467-7660.1988.tb00301

Kapteijms, L., 2013. *Clan cleansing in Somalia: the ruinous legacy of 1991*. Philadelphia: University of Pennsylvania Press.

King, C., 2001. The benefits of ethnic war: understanding Eurasia's unrecognized states. *World Politics*, 53 (4), 524–552. doi:10.1353/wp.2001.0017

Krohley, N., 2015. *The death of the Mehdi army: the rise, fall, and revival of Iraq's most powerful militia*. London: Hurst & Company.

Lacher, W., 2012. *Organized crime and conflict in the Sahel-Sahara region*. New York: Carnegie Papers.

Lecocq, B., 2010. *Disputed destert: decolonization, competing nationalisms and Tuareg rebellion*. Leiden: Brill.

Mampilly, Z., 2015. *Rebel rulers: insurgent governance and civilian life during war*. Ithaca, NY: Cornell University Press.

Markakis, J., 1987. *National and class conflict in the Horn of Africa*. New York: Cambridge University Press.

McGowan, P., 2003. African military coups d'état, 1956–2001: frequency, trends and distribution. *Journal of Modern African Studies*, 41 (4), 339–370. doi:10.1017/S0022278X0300435

Miller, N., 1981. *The other Somalia: illicit trade and the hidden economy*. Hannover, NH: American Universities Field Staff.

Munson, P., 2009. *Iraq in transition: the legacy of dictatorship and the prospects for democracy*. Washington, DC: Potomac Books.

North, D., *et al.*, 2007. Limited access orders in the developing world: a new approach to the problems of development. *World Bank Research Working Paper no. 4359*. doi:10.1094/PDIS-91-4-0467B

Patrick, S., 2007. 'Failed' states and global security: empirical questions and policy dilemmas. *International Studies Review*, 9 (3), 644–662. doi:10.1111/j.1468-2486.2007.00728.x

Peters, D., *et al.*, 1999. *Health expenditures, services and outcomes in Africa*. Washington, DC: World Bank.

Pollack, K., 2002. *The threatening storm: the case for invading Iraq*. New York: Random House.

Porch, D., 2013. *Counterinsurgency: exposing the myths of the new way of war*. New York: Cambridge University Press.

Rigaud, F., 2001. Irak: le temps suspendu de l'embargo. *Critique Internationale*. 11, 15–24. doi:10.3917/crii.011.0015

Roessler, P., 2011. The enemy within: personal rule, coups and civil war in Africa. *World Politics*, 63 (2), 300–346. doi:10.1017/S0043887111000049

Roy, O., 2018. Introduction. *In*: V. Collombier and O. Roy, eds. *Tribes and global Jihadism*. London: Hurst & Company, 1–15.

Sambanis, N., 2001. Do ethnic and nonethnic civil wars have the same causes? *Journal of Conflict Resolution*, 45 (3), 259–282. doi:10.1117/0022002701045003001

Sarkees, M., 2010. *The list of non-state wars*. Correlates of War Data Set. State College, PA: Pennsylvania State University.

Scheele, J., 2009. Tribus, États et fraude: la région frontalière algéro-malienne. *Études rurales*, 184, 79–94. doi:10.4000/etudesrurales

Scheele, J., 2012. *Smugglers and saints of the Sahara: regional connectivity in the twentieth century*. New York: Cambridge University Press.

Shurkin, M., Pezard, S., and Zimmerman, S., 2017. *Mali's next battle: improving counterterrorism capabilities*. Santa Monica, CA: RAND.

Sluglert, P. and Farouk-Sluglert, M., 1990. *Iraq since 1958: from revolution to dictatorship*. London: IB Tauris.

Snyder, J. and Jervis, R., 1999. Civil wars and the security dilemma. *In*: B. Walter and J. Snyder, eds. *Civil wars, insurgency and intervention*. New York: Columbia University Press, 15–37.

Somali Ministry of National Planning, 1987. *Report on the first meeting of the joint coordination group for Somalia*. Mogadishu: Somali Ministry of National Planning.

Sommer, J., 1994. *Hope restored? Humanitarian aid in Somalia, 1990–1994*. Washington, DC: Refugee Policy Group.

Sweet, R., 2018. *Bureaucrats at war: public authority in Congo's conflict zones*. (PhD) dissertation. Northwestern University.

Tull, D. and Mehler, A., 2005. The hidden costs of power-sharing: reproducing insurgent violence in Africa. *African Affairs*, 104 (416), 375–398. doi:10.1093/afraf/adi034

Vlassenroot, K. and Raeymaekers, T., 2004. The politics of rebellion and intervention in Ituri: the emergence of a new political complex. *African Affairs*, 103 (412), 385–412. doi:10.1093/afraf/adh066

Weinstein, J., 2007. *Inside rebellion: the politics of insurgent violence*. New York: Cambridge University Press. .

Williams, P., 2016. *War & conflict in Africa*. Cambridge: Polity.

Wood, E., 2008. The social processes of civil war: the wartime transformation of social networks. *Annual Review of Political Science*, 11, 539–561. doi:10.1146/annurev.polisci.8.082103.104832

World Bank, 1991. *Somalia: crisis in public expenditures management*. Washington, DC: World Bank.

Wucherpfennig, J., *et al.*, 2012. Ethnicity, the state and the duration of civil war. *World Politics*, 64 (1), 79–115. doi:10.1017/S004388711100030

Comparing Conflict-related Sexual Violence: Expertise, Politics and Documentation

Róisín Read

ABSTRACT

This paper explores knowledge production about sexual violence and highlights the role that comparison, across contexts and between different kinds of knowledge, plays in shaping what we know. It explores the role of strategic knowledge in creating thematic areas of international expertise and draws attention to the roles quantitative evidence and narrative testimony evidence have played in this field of knowledge. The paper addresses the role that comparison plays in producing both quantitative and qualitative data about sexual violence in conflict and considers the functional similarities in the ways these data are deployed to support sexual violence campaigning and advocacy.

Introduction

The central goal of this paper is to consider: how we know what we know about conflict-related sexual violence and to ask what role comparison might play in that knowledge. While the prompt for this special issue was to consider the politics of comparing armed conflict, the starting point for this paper was an absence of data. As part of a broader project considering the use and production of an historical sample of peacekeeping data in Darfur,[1] we noted something troubling about the data – there was no systematic attempt on the part of those compiling the data to track the impact of the conflict on women. Given the efforts of those working on the Women, Peace and Security agenda to mainstream gender in peacekeeping missions, it seemed a clear oversight that there was no attempt to disaggregate this key data on security incidents by gender. The only place that women's experiences of the conflict were visible in the data was as victims of sexual violence. This was striking: that not only was the data not being coded and stored in a way that enabled systematic analysis of the gendered nature of the conflict, but that the only place women's experiences could be easily identified, was as victims of sexual violence, and even that was much lower than expected.

Investigating this puzzle led to an exploration of the vast and rich literature on sexual and gender-based violence in conflict. It also led to an interrogation of the literature on data and knowledge production practices in global governance, and especially in conflict. The modest contribution that this paper seeks to make is in bringing these literatures into conversation around the specific question of the politics of comparing conflict-related sexual violence across contexts. In doing this, the paper makes three key contributions, the first is that the establishment of conflict-related sexual violence as an area of international public policy expertise relies upon the production of experts whose knowledge of the issue is transportable, and, thus, comparable, across contexts. The second contribution of the paper builds on the work of, for Boesten (2017, 2018); Skjelsbæk (2018, 2006) and Davies and True (2015, 2017) who have highlighted the apparent disconnect between epistemological approaches to knowledge about conflict-related sexual violence. They have especially highlighted disconnect between those who study it with quantitative data and those who use qualitative data. What I begin to explore is the ways in which, as part of the production of international expertise, quantitative evidence and qualitative evidence are deeply enmeshed in a field of knowledge. The third contribution of the paper is to compare some of this quantitative evidence and qualitative testimony evidence, to consider how – in serving the needs of international expertise for both of these kinds of evidence to be comparable across contexts – they function similarly to strip context and, in doing so, may inadvertently depoliticise the issue while doing important awareness raising.

The paper will proceed by considering the literature on the emergence of fields of strategic expertise in global public policy. It will highlight the role that both quantitative knowledge and narrative or testimony knowledge has played in providing 'evidence' for international policy on conflict and human rights. The paper will then review the literature on conflict-related sexual violence and highlight the importance of these two kinds of knowledge – quantitative and testimony – to our knowledge of this issue specifically. In doing so, it will begin to show how, while appearing as relatively distinct and disjointed, both exist within, and are produced by, a strategic knowledge field and are deeply enmeshed. It will do this by considering how the data for the Sexual Violence in Armed Conflict Dataset is produced. The paper will then begin to explore how both of these knowldges facilitate comparison across contexts and highlight some of the effects of this, noting the functional similarities in the deployment of 'killer facts' (Oxfam International 2017) and testimonies.

Expertise in Global Governance

There has been growing attention in recent years to the ways in which 'expert knowledge acts as a strategic resource' (Littoz-Monnet 2017, p. 584). Some analyses have focused on how 'purposive international bureaucracies' use expert knowledge as a way to increase 'their sphere of authority and to creep into entirely new issue domains' (ibid). Others have focused on the epistemic communities created around specific areas of expertise (Lemay-Hébert and Mathieu 2014) or the role of knowledge 'brokers' and intermediaries who use their expertise to further 'a mix of actors' own policy and career interests' (Kostić 2017, p. 121, see also Merry 2011, Fisher 2017). Increasingly, organisations in involved in global public policy, be it international organisations, such as the agencies of the UN, or international financial institutions or non-governmental organisations, justify their roles and existence on the basis of their expertise. From the World Bank's reinvention as 'the knowledge bank' (World Bank 1998); to strategic decisions on the part of operational NGOs, such as Oxfam, to turn more toward a knowledge 'broker' role to differentiate itself in a 'crowded market' (Ocaña 2011, pp. 1; 9), expertise is a major currency of the 'international community' and those local actors who seeks to interact with it (Merry 2006). In part this move is driven by the rise of evidence-based governance 'a broad range of regulatory strategies that rely on empiricism, quantitative knowledge as the basis for decision making' (Merry 2016b, p. 10). Litthoz-Monnet's suggests that (2017, p. 584), this is a response to a crisis of legitimacy on the part of international organisations who seek to justify their policies as '"evidence-based", "rational", and founded on neutral expertise'. In this way it is a fundamentally de-politicising move, taking authority for decision-making from the realm of politics and placing it in the technical/bureaucratic realm.

This has been observed specifically in relation to knowledge about conflict, with recent scholarship noting that think tanks and policy organisations can operate as conflict experts (Bliesemann de Guevara 2014, p. 546). These organisations, and the experts they employ, are deeply implicated in competition for conflict knowledge authority. There are incentive structures at work for these experts; they operate within 'knowledge markets' (Bliesemann de Guevara 2014, p. 549). The language of markets is helpful, as it encourages us to consider the competitive dynamics which drive the production and circulation of this expertise (Beerli 2017). The currency of expertise is not just confined to international organisations and non-governmental organisations, but is a complex field which includes think tanks, advocacy organisations and universities who compete for resources and visibility in a highly populated field of experts, or epistemic community. Lemay-Hébert and Mathieu (2014, p. 233), highlight the importance of thinking about '*internal* processes leading to the creation of knowledge' within epistemic communities on conflict that operate 'as the sites of a constant struggle over how to define what qualifies as valid knowledge'.

The struggle over defining valid knowledge is important to consider as it is the 'social life' of data in global governance which transforms it into 'facts' (Leite and Mutlu 2017). The enthusiasm for evidence-based policy has meant a limited vision of 'evidence' has been promoted. The focus on the rational and objective has meant quantitative forms of knowledge with their apparent facticity have tended to be preferred. Statistics have always been about establishing control, originally for the state, but increasingly for institutions and organisations of global governance (de Siqueira *et al.* 2017, p. 27, for more on this with reference to sexual exploitation and abuse in the UN see, Grady 2016). Quantification is a key move that has been observes with regards to the development of UN 'expert' knowledge (Bueger 2015, p. 9). Yet, it is important to remember that 'raw data is an oxymoron' (Rosenberg 2013), as 'whether consciously or unconsciously, [statistics] are always produced, processed, and consumed with a social objective or purpose in mind' (Leite and Mutlu 2017, p. 73). Thus, the social objective of any particular use of data is key to consider, as the decision of what to count is a crucial aspect of 'the power of quantification to shape public knowledge' (Merry 2016a, p. 107). This must analysed alongside the 'social life' of that data: the 'twin processes of data collection and analysis' (Leite and Mutlu 2017, p. 74).

While the 'seductions of quantification' (Merry 2016b) in global governance have been critiqued recently, it is by no means the only type of knowledge that is valued. Alongside the emergence of this data-driven regime of strategic expertise we have also seen a rise in narrative forms of knowledge. Fernandes notes 'an emergent culture of storytelling that presents carefully curated narratives with predetermined storylines as a tool of philanthropy, state craft and advocacy' (Fernandes 2017, p. 2). In particular, in the case of human rights abuses, the use of life narratives and testimonies has become common place, both for evidentiary 'fact-finding' and for advocacy, campaigning and fundraising (Schaffer and Smith 2004).

While storytelling as a consciousness-raising strategy emerged in indigenous movements in Latin America and US feminist movements in the 1970s, this strategy has been re-purposed over time to serve state, international organisation and nongovernmental organisation interests. Truth and reconciliation commissions and the testimony they presented ware a part of this; rather than antagonistic politics of transformation, the sharing of stories in these venues was used as a form of individual therapeutic healing which allowed for the avoidance of structural violence (Fernandes 2017, p. 17). Within this knowledge market, '[s]tories were shorn of their nuance and complexity to become short texts that would fit in a report or could be easily recitable for the purposes of a legal hearing' (Fernandes 2017, p. 17). This mode of storytelling has been central to the rise of human rights; the individual narratives told 'provide necessary evidence and information about violations' (Schaffer and Smith 2004, p. 3). Yet, while storytelling claims to give a voice to the voiceless, too little attention has been paid to the political economies within which these stories are produced and how this shapes the stories that get told and determines those that do not (Kindersley 2015).

This section has explored the rise in expertise in global governance, noting the role of evidence-based policy, and its prioritisation of quantitative ways of knowing. Key to both these kinds of knowledge being transformed into issue-specific expertise is comparison. The idea of international expertise on a specific issue relies on the assumption that knowledge gained in one context can be applied in another context. Experts facilitate the transfer of expertise from one country to another, creating space for thematic or technical expertise at the international level. The section has also highlighted the importance of a greater interrogation of data production and its relationship to this international expertise, especially a need to look at the social life of data.

Alongside the rise in quantitative evidence, it also drew attention to narrative forms of 'evidence'. Narrative evidence has played a key role human rights campaigning, but some have questioned its claims to 'give voice' by highlighting the specifics of their production. Both quantitative and narrative forms of evidence, then, are produced in particular epistemic communities for particular purposes and to understand them, we need to look at their conditions of production. The following section will consider this by looking at the production of knowledge and expertise in relation to a specific issue in international politics: conflict-related sexual violence. It will ask how we know what we know about conflict-related sexual violence and will explore one particular site of knowledge production – the Sexual Violence in Armed Conflict Dataset – to begin to provide an answer.

How Do We Know What We Know about Conflict Related Sexual Violence?

The previous section drew attention to two kinds of knowledge production which have been central to the emergence of international 'experts' in public policy. What then, are the implications of the dominance of these two kinds of knowledge production for knowledge about conflict-related sexual violence? Campbell (2018, p. 470) suggests that it is necessary for us to analyse 'research on conflict related sexual violence as a "field of knowledge"', which is rapidly expanding. Boesten (2017, p. 507) notes a particular challenge to this field of knowledge;

> Qualitative research that emphasizes local contexts and histories, as well as the specific features of gender inequality, may allow for more complex understandings of sexual violence but does not easily support policy making. After all, policy benefits from measurable indicators and generalizable definitions; contemporary evidence-based governance demands quantitative knowledge

This is often presented as a binary choice, between detailed contextual knowledge and more 'useful' quantitative knowledge. Yet, as we will see, testimony evidence, context specific knowledge and quantitative knowledge are entangled in this field of knowledge. Testimony evidence is prevalent, especially in reports by human rights and non-governmental organisations and these reports form the basis of the quantitative data on conflict-related sexual violence. This section will consider the production of the Sexual Violence in Armed Conflict dataset in order to explore this field of knowledge and in thinking about the 'social life of data' we will see the connections between narrative forms of knowledge and quantitative knowledge.

Documentation and Quantification

Attention to sexual and gender-based violence in conflict has never been higher, as highlighted by the 2014 Global Summit to End Sexual Violence in Conflict, and subsequent campaigning by the UN and others[2] on the issue. Problems of sexual violence in conflict are currently high in the rhetoric of international debates, with high profile summits and celebrity activists garnering much media attention for the issue. This is a victory for campaigners who have sought to have violence that is seen to overwhelmingly affect women in conflict taken seriously.[3] The quantification of sexual violence in conflict has been a key part of such efforts; highlighting the scale of the problem has been an important step in putting the problem on international policy-making agendas. As Cole (2018, p. 3) notes, there has been 'a trend towards investigation through data gathering' in relation to sexual violence in conflict from the 1990s onward. The establishment of sexual violence in conflict as an issue on the international agenda is commonly traced to the scale of sexual violence in the Rwandan and Yugoslavian wars in the 1990s, which 'made it politically impossible for others to supress discussion' of the issue (Kirby 2015, p. 459). The scale of mass atrocities in these conflicts meant that it was no longer possible to maintain the situation where 'silence and inaction' were the main responses of the international community (Aroussi 2011, p. 577).

Following much campaigning to have the issue recognised, in 2000, UN Security Council resolution 1325 (UNSC 2000) officially recognised sexual violence in conflict as a specific security problem.[4] The resolution made addressing sexual violence a priority, calling on 'all parties to armed conflict to take special measures to protect women and girls[5] from gender-based violence, particularly rape and other forms of sexual abuse, and all other forms of violence in situations of armed conflict' (UNSC 2000, p. 3). Documentation is a central concern in UN resolution 1325, with the recognition of the 'need to consolidate data on the impact of armed conflict on women and girls' (2000, p. 2); a commitment followed up in subsequent resolutions with specific requests for data on sexual violence in situations of armed conflict (UNSC 2008, p. 4, 2009a, p. 7).

As Bernard and Durham (2014, p. 427) note in their introduction to a special issue of the *International Review of the Red Cross* on sexual violence in armed conflict; 'Sexual violence has been, and to a large extent continues to be, shrouded in silence'. Prior to UN resolution 1325, Nordstrom (1999, p. 71) argued, in relation specifically to violence against girls in conflict, that silence, manifested as a lack of statistical and other documentation is a political choice:

> Girls constitute invisible casualties of war ... it is an invisibility that takes on cultural dynamics when a dearth of statistics, accounts, and accountability perpetuate a tradition of "not-seeing" these forms of violence in the world. If silence is political, "not-knowing" is at the core of power and its abuses.

The documentation and quantification of sexual violence in conflict can, therefore, be read as a political refusal of this silence. Documentation becomes more than a bureaucratic exercise, it is a political act, a refusal of the previous regimes of 'not-knowing'. In this context, with the recognition that the issue had been previously overlook, conflict-related sexual violence 'became increasingly front and centre stage in war reporting, fact – finding, and policy making in the 2000s' (Skjelsbæk 2018).

Yet, despite more interest in the political recognition of sexual violence it 'remains extremely difficult to quantify' (Bernard and Durham 2014, p. 427). In order to document the scale of the problem of sexual violence in conflict, it is necessary to count it. Politically, 'if it is not counted it does not count' (Andreas and Greenhill 2010, p. 1). Gender concerns have, historically, not (been) counted. For example, in academic and policy arenas there is a lack of sex and age disaggregated data (Mazurana *et al.* 2013). Many of the major academic conflict datasets have fatality requirements for an event to be included, and thus, they do not necessarily capture sexual violence. In relation to attacks on aid workers, Wille and Fast (2011) observe that a key obstacle to understanding the gendered nature of security events is that 'information on gender is not systematically available'. Thus, much of the data about conflict overlooks gender as a variable of interest, though this is increasingly being recognised and steps taken to address it (for example, a commitment following the World Humanitarian summit was to collect sex- and age- disagregated data, United Nations 2016, p. 10).

Creating Quantitative Data about Sexual Violence in Conflict in a Field of Knowledge

One such effort is the Sexual Violence in Armed Conflict Dataset (Cohen and Nordås 2014). The dataset covers the years 1989–2009 and 129 active conflicts during this period. It includes 625 armed actors involved in the identified conflicts. In order to produce its data, the SVAC dataset draws on the three 'most commonly used sources in quantitative human rights scholarship: the US State Department, Amnesty International, and Human Rights Watch' (Cohen and Nordås 2014, p. 421). It is on this – the relationship between the data and the reports that underpin it – that this section will focus. In particular, it will focus on the reports from the non-governmental sources: Amnesty International and Human Rights Watch. It is not my goal here to condemn the SVAC for this, but merely to draw attention to the ways in which different kinds of knowledge combine and intertwine to produce international expertise about sexual violence in conflict and to consider what the implications of this international expertise might be. The SVAC creators note that they draw on these three sources because they are global; are considered trustworthy sources on human rights violations; and because they are used by other datasets (Cohen and Nordås 2014, p. 421). Their global perspective and their use by other datasets is important for comparison, as the next section will examine, but their trustworthiness on human rights is worth examining in more detail as it is this knowledge claim that affirms their international expertise.

The SVAC creators note the possible biases with drawing on these reports: underreporting of sexual violence; an interest in 'special categories of victims' (ibid); the possible reasons for reporting of sexual violence going up in wartime (for strategic reasons or due to a greater service provision); and that increased attention globally might result in improved reporting. The codebook also notes that other sources were consulted but not systematically coded, such as International Crisis Group reports. Organisations, such as Amnesty International, Human Rights Watch and the International Crisis Group function as 'highly visible, vocal and hard to ignore conflict expert[s]' (Bliesemann de Guevara 2014, p. 546). The reports they produce are key, as we have seen, to the production of other forms of knowledge about conflict and human rights. Yet, reports are an understudied aspect of international knowledge regimes (Bornstein 2019) and '[a]t both the institutional and the textual level, there is typically little transparency about the methodology used and little formal accountability for the work produced' (Bruch 2013).

In highlighting this, I am re-enforcing Davis and True's claim that the politicised environment in which reporting and data collection on sexual violence in conflict take place has been overlooked (Davies and True 2015, p. 507). In order to illustrate the effect this politicised environment can have on what data gets produced about sexual violence in conflict, I want to return to the absence of data that was the impetus for the research of this paper. When looking into a historical sample of data from the UN/AU hybrid peacekeeping mission in Sudan (UNAMID), it became apparent that the Joint Mission Analysis Centre (JMAC), tasked with compiling and providing real-time and strategic analysis of the conflict in Darfur, was not systematically tracking the impact of the conflict on women. Not only was there no attempt to disaggregate the data on security incidents by gender,[6] but when the security incident data was compared with interview data from Darfurian refugees, who had been in Darfur during that time, there was a mismatch between their perception of the frequency of incidents involving violence, particularly sexual violence, against women. Interviewees described sexual violence as a 'daily' problem for women.[7] The only place that women's experiences of the conflict were visible in the data was in the case of incidences of sexual violence. Of the 2861 incidents recorded in the dataset, 13 were coded as rape[8] and nine as 'GBV' [gender-based violence]. On reviewing these coded, two of the 13 were miscoding errors and of those coded as GBV, all were actually reports of sexual violence against women.

It was surprising that, given the media attention on sexual violence, and the impact of the conflict in Darfur on women more generally, the data did not seem to match. And, that the way the data had been structured, did not seem to be particularly concerned with analysing this aspect of the conflict. Though I was not able to speak to the dataset's producers, I was able to interview someone who worked in the Human Right section of the mission several years later.[9] They were not surprised at the lack of this data in the dataset, but claimed that more comprehensive data did exist within the Human Rights section. When I asked why it was not shared with the Joint Mission Analysis Centre, the interviewee claimed that, though there were concerns about the confidentially of the data in the Mission – which has been highlighted as a problem in the mission (Read 2016) – perhaps a bigger part of it was that 'our data was our currency' (interview with author, 2015). By this, he explained, that there was a lot of international attention to the issue of sexual violence in conflict and that by having control and ownership over the data, the Human Rights section – within Darfur and at the head office they reported to in Geneva – were able to ensure their place at the table for decision-making over policy, at the local and international level. This seems to reflect the literature on international expertise and knowledge, or in this case data, as a strategic resource (Littoz-Monnet 2017). Though it is not possible to draw a general inference from one perspective on one case of data collection about conflict-related sexual violence, it points to the need for future research into the politics of knowledge production and expertise in this area and to consider what impact this might have on who has access to knowledge produced and how it is used to further institutional agendas in international organisations.

In thinking about the example of the missing data from the Darfur case, we can see that institutional politics might play a role, but so too does the broader national political environment. Returning to the Amnesty International and Human Rights Watch reports that form the basis of the SVAC dataset, it is worth considering the extent to which the political environment they are produced in impacts on their findings. Chaudhry (2019, p. 1) notes that this is an area where much more research is needed, noting our lack of knowledge about 'the data-generating processes that yield these [INGO human rights] reports'. Her study uses a sub-national analysis to explore variation across Amnesty International, Human Rights Watch and a local rights monitoring organisation in India. Based on her findings, she argues that 'there is clear evidence that NGOs are strategic in reporting and undertaking their naming and shaming activities' (Chaudhry 2019, p. 3). Additionally, the 'duelling incentives' in the information politics of transnational advocacy organisations is something that the creators of the SVAC are aware (Cohen and Green 2012).

Understanding the political context in which data collection and reporting on sexual violence in conflict occurs matters for considering the extent to which these, apparently objective reports are used to produce apparently objective statistics. In the production of the SVAC, the coding of the prevalence of sexual violence rests on the use of particular terms. A prevalence score of three (the highest) is given where the reports use the descriptions: '"massive," "innumerable," or "systematic"' or where the actor using sexual violence is described as using it 'a "means of intimidation," "instrument of control and punishment," "weapon," "tactic to terrorize the population," "terror tactic," "tool of war," on a "massive scale"' (Cohen and Nordås 2013, p. 10). Others have written about the implications of framing sexual violence in as a weapon (Crawford 2013, 2017), or the importance of terms like 'systematic', in establishing conflict-related sexual violence as 'exceptional, different and not inevitable' (Boesten 2017, p. 507). The key point here the co-constitutive nature of this knowledge, documenting the scale of the phenomena in conflict settings is a key move in demonstrating that conflict-related sexual violence is 'systematic',[10] but the descriptors of that are established by this counting are also then used as proxies for determining the levels. Indeed, a prevalence score of two is given when words like 'widespread', 'pattern' or 'numerous' appear, but if the numbers given are in the prevalence score two range, but the word 'systematic' is given then the instruction is to code based on the text description at the higher three level. The issue here that arises from this relationship between the narrative evidence presented in the reports and the quantitative evidence produced using those reports is that the SVAC aims to highlight the scale of the problem of sexual violence in conflict and to use that quantitative evidence to ty and understand the nature of the problem. But their quantification of the scale of the problem is already influenced by the reports' analysis of the nature of the problem. Again, my goal here is not to criticise the SVAC for this, but to point to the importance given to the reports and the international expertise they appear to represent.

It is by no means a novel point to note that definitions for coding decisions matter and influence analysis. Campbell points to this problem more broadly in the knowledge field about conflict-related sexual violence in defining the object of study (Campbell 2018). Others still have pointed to the challenges and difficulties with establishing shared definitions of conflict-related sexual violence or even of sexual violence at all (for some examples, see, Auchter 2017, Baaz and Stern 2018, Campbell 2018). The point here is not to offer a 'better' or all-encompassing problem free-definition, but to emphasise the role that definitions play in shaping our knowledge about conflict-related sexual violence. As the next section will explore in more detail, the fixing of definitions is important aspect to consider in understanding how the desire to compare knowledge about conflict related sexual violence shapes that knowledge.

How Does Comparison Fit In?

A key way to facilitate understanding and analysis of data about conflict-related sexual violence is comparison. In their introduction to the sexual violence in armed conflict dataset, Cohen and Nordas (2014, p. 418), note that most quantitative studies of conflict-related sexual violence are 'case studies of specific conflicts' and are 'not comparative in nature'; and, following Wood (2006), do not explore cases where sexual violence has not taken place. They argue that systemic comparison could 'illuminate causal mechanisms and root causes' which could be crucial to developing effective policy initiatives and addressing the effects of conflict-related sexual violence (Cohen and Nordås 2014, p. 418). Comparison, here, is presented as a key feature of understanding the occurrence (and non-occurrence) of conflict-related sexual violence. The section will consider how conflict-related sexual violence has been compared across contexts, both geographic and temporal.

Comparison relies on us making things comparable and central to this is the development of definitions. The absence of standardised definitions of sexual violence, historically, has prevented the 'development of effective reporting systems and/or databases and thus restrains prevention, monitoring and victim advocacy efforts' (Dewey and St Germain 2012, p. 51). The development of definitions in relation to quantification is key to comparison. Standard definitions allow for comparison not only across geographic contexts but also over time, eventually naturalising such classifications and categorisations (de Siqueira *et al.* 2017, p. 27). My goal here is not to get into debates about what should or should not be included in definitions of conflict-related sexual violence, but to consider how the fixing or not of definitions enables particular forms of comparison and hinders others. In order to do this, I will look at some examples from the Sexual Violence in Armed Conflict dataset's definitions as set out in their codebook. Setting clear definitional rules is necessary in facilitating comparison, the point here is not to claim that these are not carefully considered issues in the production of the dataset, but to highlight what the limitations might mean, they enable comparison but they also limit our understanding in specific ways.

In order to facilitate comparison of conflict-related sexual violence of the kind that Cohen and Nordås highlight as essential, to 'illuminate causal mechanisms and root causes', (Cohen and Nordås 2014, p. 418), it is necessary to fix definitions. A key aspect of this definitional work of note is way that the SVAC defines sexual violence as conflict-related if it is carried out by conflict actors:

> *Perpetrators* include only organized armed actor groups, the main protagonists of armed conflicts – state forces, rebel groups or PGMs [pro-government militias]. We do not code perpetrator data at the individual level (e.g. a particular commander in an army), nor domestic violence, intimate partner violence, violence by peacekeepers or aid workers, or violence by civilians (Dara Kay Cohen and Nordås 2014, p. 419)

There are clear and logical reasons for this; that the data does not currently exist at the requisite level of detail for those aspects not coded. However, this impacts the nature of the comparison that can be carried out and the state of our knowledge about sexual violence in conflict.

Sexual violence carried out by peacekeepers and humanitarians operating in conflict contexts is an illustrative example for thinking about how definitions (and their exclusions) shape the possibilities for comparison (Smith 2018). This sexual violence could be viewed through the lens of the conflict, certainly in peacekeeping missions, such as the one in Darfur, there would be an argument for viewing peacekeepers as conflict actors. The effect of this exclusion is to create an exception. The exclusion of violence committed by civilians, for example, also impacts on the nature of the comparisons that can be conducted, it leads to understanding conflict-related sexual violence as a military phenomenon. This can reinforce understandings of sexual violence in conflict as a systematic weapon of war, rather than, for example 'driven by runaway norms that legitimize rape, torture, and other atrocities' (Leatherman 2017), leading to 'a separation in studies of [conflict-related sexual violence] on the one hand and of peacetime violence on the other' (Boesten 2018, p. 458). This makes a comparison between pre- and post-conflict and conflict related sexual violence, which some have highlighted as a key feature of understanding the occurrence of sexual violence in conflict (Butler and Jones 2016).

The need for definitions to be standardised to facilitate comparison is important when thinking about how a field of knowledge is built. As has already been noted, conflict datasets are very reliant on a small number of 'conflict experts', such as Amnesty International and Human Rights Watch, and 'the high level of credibility and authority these reports have gained, far outweighing any similar studies by governmental or UN bodies' (Dudai 2007). But we must recognise that while this has advantages for comparison, it also limits the diversity of knowledge. If definitions are increasingly standardised and, 'designed to be compatible with other relevant and widely used datasets' (Cohen and Nordås 2014, p. 425) and the information that populates those definitions is drawn from limited range of sources, this narrows what we can know about conflict-related sexual violence. In this way, the establishment of conventions of counting with a broad range of international support matters. The support for schemes of measurement is what provides them with their authority. They are 'collective product[s]' which have little standing until 'authorised by disciplinary specialists' (Loughlin 2003), reinforcing the role that existing expertise at the international level plays in legitimating the information on which that claim to expertise is based.

Sally E Merry argues that a key political effects of the need to make things comparable in order to count them is 'that they are inevitably stripped of their context, history, and meaning' (Merry 2016a, p. 1). In many ways this is a trade of between different perspectives on a phenomena, in order to be able to see patterns across time, space and conflict actors, you need to sacrifice some context-specificity to facilitate comparison. In thinking about the international recording and monitoring of sexual violence in conflict, standardisation of both definitions and measurement is crucial to allow for quantitative comparison across contexts. It is this that helps facilitate the establishment of it as an area of global public policy; '[t]he politics of numbers is part of the media frenzy and hyper response' to sexual violence in conflict (Hilhorst and Douma 2018, p. S87). Yet, earlier sections also noted the role of narrative and testimony evidence which human rights reporting draws on. In this final section, I will argue that, just as the need to make things comparable in order to count them strips context, history and meaning, so too does the production of testimony and narrative for the purpose of human rights, and sexual violence, monitoring across contexts.

Returning to the case of Sudan, Kindersley (2015, p. 204) notes that, '[m] edia, aid, and academic work have struggled for decades with clearly articulating events in Sudan beyond a massive ahistorical and apolitical numbers game'. As a response to this, and other similar needs to make complex crises knowable, NGOs turned to storytelling to in an attempt 'to make large-scale crises comprehensible and translatable' (ibid). We can see this in human rights work more generally (Schaffer and Smith 2004, Bake and Zöhrer 2017). However, I will argue in this section, that narrative and testimony about sexual violence drawn from human rights reporting is often deployed in a similar way to statistical evidence such as 'killer facts' (Oxfam International 2017). That, in needing to make the experiences of conflict-related sexual violence comprehensible and translatable across contexts, there is – as we also saw with quantitative forms of knowledge – a loss of context, history and politics. In order to demonstrate this, I analyse the 'testimony' section of the UN Action Against Sexual Violence in Conflict (UN Action) website (www.stoprapenow.org).[11]

UN Action, formed in 2008 by the passing of UNSCR 1820, brought together 13 agencies in the UN to address the issue of sexual violence in conflict. The website features seven key sections, highlighted in tabs at the top of the page: Take Action; Field Updates; News; Testimonies; NGO Links; About; Advocacy Resources (UN Action, n.d.). The 'Testimony' section features 12 short personal testimony extracts drawn from a range of international reports, news sources and women's organisations. Of the twelve, three are from Human Rights Watch and two from Amnesty International Reports.[12] While most of the extracts are from between 2003–2007, one is from 1995. All the extracts appear in individual boxes, which include the extract and the source, against a backdrop of equally sized square grey pictures of individuals with their arms crossed (presumably from the 'Get Cross' campaign the website ran asking people to submit pictures indicating their support for UN Action's work). The header image is a colour photo of just the arms crossed over the chest (with the head and body cropped out). Both the images and testimony appear on this page without any additional information to explain or contextualise them. There is no information given about why these extracts were chosen or what the section aims to do. In the absence of other information, the section appears to function as a global overview of the issue of sexual violence in conflict, in providing a sample of extracts from which we can compare and see similarities and differences across a selection of contexts. Yet, these, in some cases extremely, graphic testimonies of sexual violence have been removed from the Human Rights Watch and Amnesty International Reports which provided the history and context to make this violence meaningful in specific places and specific conflicts. Instead, they simply appear as horror stories designed to shock and provoke sympathy but not enhance understanding.

In this way, the use of narrative testimony, far from serving to provide 'local contexts and histories, as well as the specific features of gender inequality, [which] may allow for more complex understandings of sexual violence' (Boesten 2017, p. 507), actually functions to strip history and context and so limits understanding. This has similarities with the use of 'killer facts' in advocacy and campaigning; 'those punchy, memorable, headline-grabbing statistics that make reports special. They cut through the technicalities to fire people up about changing the world. They are picked up and repeated endlessly by the media and politicians' (Oxfam International 2017, p. 1). One of the keys to a good 'killer fact' is 'some kind of comparison' (ibid). Oxfam's From Poverty to Power blog (Green 2015) links to an article in *The Telegraph* to give an example of the use of killer fact about sexual violence in conflict. The article has the headline 'Woman raped in Congo "nearly every minute"' and the 'killer fact' it draws on is that 'A woman in certain parts of the Congo is 134 times more likely to be raped than a woman in the United States' (Pflanz 2011). Just as testimony serves to put a human face of an issue such as sexual violence in conflict, and foster an immediate emotions reaction, these statistics are designed to provoke an emotional reaction, rather than to encourage detailed understanding. Yet, both are 'picked up' by others and used in campaigning and advocacy.

My goal is not to say that these are not valid and effective campaigning tools but to point out that, in the creation of a sphere of global public policy expertise around sexual violence in conflict, there are similarities in the ways that testimony evidence and quantitative evidence are deployed, in that both can function to flatten and strip context. In thinking about how we know what we know about conflict-related sexual violence, and the role that comparison plays in this, it is import to pay attention to the similarities and to further consider what the effects of them might be on the types of policies and campaigns that are enacted. While Amnesty International's annual reports have been designed specifically to be non-comparative in nature (Dudai 2007) and their country or issue specific reports, as with those of Human Rights Watch, seek to provide detailed, context-specific information, this paper has highlighted the need to consider the role they play in the 'field of knowledge' on conflict related sexual violence. In particular, the way that they are foundational sources of knowledge for forms of comparison of the issue of sexual violence in conflict globally, whether in an abstracted focus on the testimony evidence they provide or in the way that their information is used to produce quantitative data.

Conclusion

This paper has sought to explore how we know what we know about sexual violence and to highlight the role that comparison, across contexts and between different kinds of knowledge, plays in shaping what we know. It began by highlighting the role of strategic knowledge in facilitating the creation of thematic areas of international expertise, such as that about sexual violence in conflict. It then drew attention to the role that quantitative evidence and narrative testimony evidence have played in this field of knowledge. The paper then sought to consider the social life of data and to draw attention to 'twin processes of data collection and analysis' (Leite and Mutlu 2017, p. 74) in which quantitative and qualitative forms of evidence cannot be seen as separate, given the extent to which they are enmeshed. Finally, the paper addressed some of the roles that comparison plays in producing both quantitative and qualitative data separated from meaningful context and analysis of the politics of sexual violence in conflict. It highlighted the need to consider the functional similarities between the ways data is deployed to support campaigning and advocacy and called for further research into the social life of the field of expertise around conflict-related sexual violence. In particular, questions remain about actual practices of data production and interconnectedness of these apparently diverse sources of knowledge.

Notes

1. The data is a historical sample of data from the Joint Mission Analysis Centre (JMAC) of the UN/AU Hybrid mission in Darfur (UNAMID). It covers January 2008-April 2009. JMACs are responsible for 'integrated operations monitoring, reporting and information analysis hubs at Mission headquarters to support the more effective integration of mission-wide situational awareness, security information and analysis for management decision-making' (UN DPKO, 2006, p. 2).
2. For example the Preventing Sexual Violence Initiative, led by the UK Government under William Hague https://www.gov.uk/government/organisations/preventing-sexual-violence-initiative .
3. Though some have challenged the view that that 'women and girls are "overwhelmingly" the victims of sexual violence' (Kirby 2015, p. 496), it is still the dominant framing of sexual violence.
4. Subsequently a number of other UNSC resolutions have addressed this issue, 1820 (2008); (2009a): 1888 (2009b) (2009); 1889 (2009c); 1960 (2010); 2106 (2013); 2242 (2014); .
5. The focus on 'women and girls' as the predominant victims of sexual violence is persistent across all the policy documents. For example, the issue continues to be championed as part of the Women, Peace and Security Agenda and Human Rights Watch includes sexual violence under the Women's Rights category on its website.
6. As has been observed, this is by no means a problem limited to the UNAMID mission.
7. 53 interviews with Darfurian refugees in Chad were conducted between April-June 2015 by [removed for review]. Inclusion was on the basis they were in Darfur for at least some of the same period as the data covered and they were asked to reflect on their memories of this time.
8. The incidents were primary coded – meaning that if a report mentioned an attack on civilians which also featured sexual violence, that incidence would be coded only as an attack on civilians.

9. Interview with former UMAID human rights section officer, Manchester, December 2015.
10. For more on the framing of CRSV as widespread and systematic, see Eriksson Baaz and Stern (2013); Davies and True (2015), Sarah Smith (2018).
11. Though the website is still live, it does not appear to be updated and the UN Action information, and more current information, can be found on the United Nations Office of the Special Representative of the Secretary-General on Sexual Violence in Conflict Office of the Special Representative of the Secretary-General on Sexual Violence in Conflict section of the UN website (https://www.un.org/sexualviolenceinconflict/). However, the testimonies that appear in the section are drawn from international human rights reports, including from Amnesty International and Human Rights Watch, covering at least some of the same time period as the data of the SVAC which also drew on these source. Others have analysed the content of the 'Stop Rape Now' website, pointing out its obfuscation of male victims (Grey and Shepherd 2013).
12. All the other sources have only one extract except a 2007 article in the Christian Science Monitor from which there are two extracts and actually one of these is repeated but attributed to a different source in error.

Acknowledgments

I would like to thank the editors Bruno Charbonneau and Adam Sandor for inviting me to the initial workshop and for all their subsequent support and feedback. Thanks as well to all the other participants at the *Politics of Comparing Armed Conflict* workshop for their thoughtful comments and input. I would also like to thank Sarah Smith and Lydia Cole for the interesting conversations on this topic and kindly letting me read their works-in-progress and to the anonymous revisers for their detailed and helpful advice and input. Much thanks goes to the members of the Making Peacekeeping Data Work for the International Community team, and especially to Alex de Waal for sharing the data that made the project possible.

Disclosure statement

No potential conflict of interest was reported by the author.

Funding

This work was supported by the ESRC under grant ES/L007479/1.

ORCID

Róisín Read http://orcid.org/0000-0003-1782-3209

References

Andreas, P. and Greenhill, K.M., 2010. Introduction: the politics of numbers. *In*: P. Andreas and K.M. Greenhill, eds. *Sex, drugs, and body counts: the politics of numbers in global crime and conflict*. London: Cornell University Press, 1–22.

Aroussi, S., 2011. 'Women, peace and security': addressing accountability for wartime sexual violence. *International Feminist Journal of Politics*, 13 (4), 576–593. doi:10.1080/14616742.2011.611663

Auchter, J., 2017. Forced male circuncision: gender-based violence in Kenya. *International Affairs*, 93 (6), 1339–1356. doi:10.1093/ia/iix183

Baaz, M.E. and Stern, M., 2018. Curious erasures: the sexual in wartime sexual violence. *International Feminist Journal of Politics*, 20 (3), 295–314. doi:10.1080/14616742.2018.1459197

Bake, J. and Zöhrer, M., 2017. Telling the stories of others: claims of authenticity in human rights reporting and comics journalism. *Journal of Intervention and Statebuilding*, 11 (1), 81–97. doi:10.1080/17502977.2016.1272903

Beerli, M.J., 2017. The power to count and the stakes of counting: an inquiry into the quantified production of humanitarian insecurity. *Global Governance*, 23 (1), 57–70. doi:10.1163/19426720-02301006

Bernard, V. and Durham, H., 2014. Sexual violence in armed conflict: from breaking the silence to breaking the cycle. *International Review of the Red Cross*, 96 (894), 427–434. doi:10.1017/S1816383115000442

Bliesemann de Guevara, B., 2014. Studying the international crisis group. *Third World Quarterly*, 35 (4), 545–562. doi:10.1080/01436597.2014.924060

Boesten, J., 2017. Of exceptions and continuities: theory and methodology in research on conflict-related sexual violence. *International Feminist Journal of Politics*, 19 (4), 506–519. doi:10.1080/14616742.2017.1367950

Boesten, J., 2018. Revisiting methodologies and approaches in researching sexual violence in conflict. *Social Politics*, 25 (4), 457–468. doi:10.1093/sp/jxy035

Bornstein, E., 2019. The report : a strategy and nonprofit public good the report : a strategy and nonprofit public good. *Humanity: An International Journal of Human Rights, Humanitarianism, and Development*, 10 (1), 109–131.

Bruch, E.M., 2013. Having the last word: human rights reporting (Re)imagined through critical qualitative methodology. *Qualitative Sociology*, 36 (2), 209–225. doi:10.1007/s11133-013-9246-8

Bueger, C., 2015. Making things known: epistemic infrastructures, the United Nations and the translation of piracy. *Ceu.Hu*. 1975, 1–26. doi:10.1111/ips.12073

Butler, C.K. and Jones, J.L. (2016). Sexual violence by government security forces: are levels of sexual violence in peacetime predictive of those in civil conflict? *International Area Studies Review*. 19 210–230. Published. doi: 10.1177/2233865916632087

Campbell, K., 2018. Producing knowledge in the field of sexual violence in armed conflict research: objects, methods, politics, and gender justice methodology. *Social Politics*, 25 (4), 469–495. doi:10.1093/sp/jxy025

Chaudhry, S., 2019. Bridging the Gap: the relationship between INGO activism and human rights indicators. *Journal of Human Rights*, 1–23. doi:10.1080/14754835.2019.1579638

Cohen, D.K. and Green, A.H., 2012. Dueling incentives: sexual violence in Liberia and the politics of human rights advocacy. *Journal of Peace Research*, 49 (3), 445–458. doi:10.1177/0022343312436769

Cohen, D.K. and Nordås, R., 2013. *Sexual violence in armed conflict (SVAC) dataset codebook and user instruction guide*. 1–31. Available form: www.sexualviolencedata.org

Cohen, D.K. and Nordås, R., 2014. Sexual violence in armed conflict: introducing the SVAC dataset, 1989–2009. *Journal of Peace Research*, 51 (3), 418–428. doi:10.1177/0022343314523028

Cole, L. 2018. From quantification to counting: legal-bureaucratic recognition and the subject of wartime sexual violence. *In*: *BISA gendering international relations working group 30 year annivsary conference*, Newcastle, UK.

Crawford, K.F., 2013. From spoils to weapons: framing wartime sexual violence. *Gender & Development*, 21 (3), 505–517. doi:10.1080/13552074.2013.846622

Crawford, K.F., 2017. *Wartime Sexual Violence: from silence to condemnation of a weapon of ware*. Washington, DC: Georgetown University Press.

Davies, S.E. and True, J., 2015. Reframing conflict-related sexual and gender-based violence: bringing gender analysis back in. *Security Dialogue*, 46, 495–512. doi:10.1177/0967010615601389

Davies, S.E. and True, J., 2017. The politics of counting and reporting conflict-related sexual and gender-based violence: the case of Myanmar. *International Feminist Journal of Politics*, 19 (1), 4–21. doi:10.1080/14616742.2017.1282321

de Siqueira, I.R., Leite, C.C., and Beerli, M.J., 2017. Powered and disempowered by numbers: data issues in global governance. *Global Governance*, 23 (1), 27–30. doi:10.1163/19426720-02301003

Dewey, S. and St Germain, T., 2012. Between global fear and local bodies: toward a transnational feminist analysis of conflict-related sexual violence. *Journal of International Women's Studies*, 13 (3), 49–64.

Dudai, R., 2007. A to Z of abuses: "State of the Art" in global human rights monitoring. *Development and Change*, 38 (6), 1255–1265. doi:10.1111/j.1467-7660.2007.00450.x

Eriksson Baaz, M. and Stern, M. 2013. *Sexual violence as a weapon of war? : perceptions, prescriptions, problems in the congo and beyond*. London, Zed Books. Available from: http://www.diva-portal.org/smash/record.jsf?pid=diva2:617148 Accessed 10 December 2018.

Fernandes, S., 2017. Curated stories: the uses and misuses of storytelling. *Curated Stories: the Uses and Misuses of Storytelling*. doi:10.1093/acprof:oso/9780190618049.001.0001

Fisher, J., 2017. Reproducing remoteness? States, Internationals and the Co-constitution of aid Bunkerization' in the East African Periphery. *Journal of Intervention and Statebuilding*, 11 (1): 98–119.

Grady, K., 2016. Sex, statistics, peacekeepers and power: UN data on sexual exploitation and abuse and the quest for legal reform. *The Modern Law Review*, 79 (6), 931–960. doi:10.1111/1468-2230.12225

Green, D. 16 January 2015. Every key stat you could possibly want about humanitarianism, emergencies etc – please steal. *From poverty to power*. Available from: https://oxfamblogs.org/fp2p/every-key-stat-you-could-possibly-want-about-humanitarianism-emergencies-etc-please-steal/ Accessed 7 June 2019.

Grey, R. and Shepherd, L.J., 2013. "Stop rape now?": masculinity, responsibility, and conflict-related sexual violence. *Men and Masculinities*. doi:10.1177/1097184X12468101

Hilhorst, D. and Douma, N., 2018. Beyond the hype? The response to sexual violence in the democratic republic of the Congo in 2011 and 2014. *Disasters*, 42, S79–S98. doi:10.1111/disa.2018.42.issue-S1

Inger, S., 2006. Victim and survivor: narrated social identities of women who experienced rape during the war in Bosnia-Herzegovina. *Feminism Psychology*, 16 (4), 373–403. doi:10.1177/0959353506068746

Kindersley, N., 2015. Southern Sudanese narratives of displacement, and the ambiguity of "voice.". *History in Africa*, 42, 203–237. doi:10.1017/hia.2015.3

Kirby, P., 2015. Ending sexual violence in conflict: the preventing sexual violence initiative and its critics. *International Affairs*, 91 (3), 457–472. doi:10.1111/inta.2015.91.issue-3

Kostić, R., 2017. Shadow peacebuilders and diplomatic counterinsurgencies: informal networks, knowledge production and the art of policy-shaping. *Journal of Intervention and Statebuilding*, 11 (1), 120–139. doi:10.1080/17502977.2016.1272905

Leatherman, J. 2017. International peace research association (IPRA) SEXUAL VIOLENCE AND ARMED CONFLICT : COMPLEX DYNAMICS OF RE-VICTIMIZATION Author (s): Janie Leatherman Source. *International Journal of Peace Studies*, 12 (1) (Spring/Summer 2007), pp. *12*(1), 53–71

Leite, C.C. and Mutlu, C.E., 2017. The social life of data: the production of political facts in EU policy governance. *Global Governance*, 23 (1), 71–82. doi:10.1163/19426720-02301007

Lemay-Hébert, N. and Mathieu, X., 2014. The oecd 's discourse on fragile states: expertise and the normalisation of knowledge production. *Third World Quarterly*, 35 (2), 232–251. doi:10.1080/01436597.2014.878129

Littoz-Monnet, A., 2017. Expert knowledge as a strategic resource: international bureaucrats and the shaping of bioethical standards. *International Studies Quarterly*, 61 (3), 584–595. doi:10.1093/isq/sqx016

Loughlin, S. 2003. *Rumsfeld on looting in Iraq: "Stuff happens"*. Available from: http://edition.cnn.com/2003/US/04/11/sprj.irq.pentagon/ Accessed 11 January 2018.

Mazurana, D., Benelli, P., and Walker, P., 2013. How sex- and age-disaggregated data and gender and generational analyses can improve humanitarian response. *Disasters*, 37 (SUPPL.1), 68–82. doi:10.1111/disa.2013.37.issue-s1

Merry, S.E., 2006. Transnational human rights and local activism: mapping the middle. *American Anthropologist*, 108 (1), 38–51. doi:10.1525/aa.2006.108.issue-1

Merry, S.E., 2011. Measuring the world. *Current Anthropology*, 52 (S3), S83–S95. doi:10.1086/657241

Merry, S.E., 2016a. *Quantification: measuring human rights, gender violence, and sex trafficking*. London: university of Chicago Press.

Merry, S.E., 2016b. *The seductions of quantification: measuring human rights, gender violence, and sex trafficking*. London: Chicago University Press.

Nordstrom, C., 1999. Visible wars & invisible girls, shadow industries, and the politics of not-knowing. *Atlantis: A Wome's Studies Journal*, 1 (1), 71–84.

Ocaña, M.S.D. 2011. *A review of Oxfam joint country analysis and strategy documents*. Available from: http://www.oxfamblogs.org/fp2p/wp-content/uploads/Review-of-JCASs_final_22_12_11.pdf Accessed 7 June 2019.

Oxfam International. 2017. *Creating killer facts and graphics*. Oxford: Oxfam GB.

Pflanz, M. 12 May 2011. Woman raped in Congo "nearly every minute". *The Telegraph*.

Read, R., 2016. Tensions in un information management: security, data and human rights monitoring in Darfur, Sudan. *Journal of Human Rights Practice*, 8, 1. doi:10.1093/jhuman/huv018

Rosenberg, D., 2013. Data before the fact. *In*: L. Gitelman, ed. *Raw data is an oxymoron*. Cambridge. MA: MIT Press, 15–40.

Schaffer, K. and Smith, S., 2004. Human rights and narrated lives: the ethics of recognition. *Human Rights and Narrated Lives: the Ethics of Recognition*. doi:10.1057/9781403973665

Skjelsbæk, I., 2018. Silence breakers in war and peace: research on gender and violence with an ethics of engagement. *Social Politics*, 25 (4), 496–520. doi:10.1093/sp/jxy031

Smith, S. 2018. " ... until provoked by crisis": scandal, shame, and the politics of responsibility for sexual and gender based violence in war. *In*: *BISA gendering international relations working group 30 year annivsary conference*, Newcastle, UK.

UN Action Against Sexual Violence in Conflict. n.d.. *Stop rape now*, Available from: http://www.stoprapenow.org/ Accessed 16 June 2019.

UN DPKO. 2006. *Joint operations centres and joint mission analysis centres*. Policy Directive: POL/ 2006/3000/04

United Nations. 2016. *Agenda for humanity: annex to the report of the secretary general for the world humanitarian summit* (No. A/70/709).

UNSC. 2000. Resolution 1325, Pub. L. No. S/RES/1325

UNSC. 2008. Resolution 1820, Pub. L. No. S/RES/1820

UNSC. 2009a. Resolution 1880, Pub. L. No. S/RES/1880

UNSC. 2009b. Resolution 1888, Pub. L. No. S/RES/1888

UNSC. 2009c. Resolution 1889, Pub. L. No. S/RES/1889

UNSC. 2010. Resolution 1960, Pub. L. No. S/RES/1960

UNSC. 2013. Resolution 2106, Pub. L. No. S/RES/2106

UNSC. 2014. Resolution 2242, Pub. L. No. S/RES/2242

Wille, C., and Fast, L. 2011. *Security facts for humanitarian agencies: Aid, gender and security: The gendered nature of security events affecting aid workers and aid delivery. Insecurity Insights*. Geneva. *November*. Available from: http://insecurityinsight.org/files/Security%20Facts%202%20Gender.pdf Accessed 10 December 2018.

Wood, E.J., 2006. Variation in sexual violence during war. *Politics & Society*, 34, 307–342. doi:10.1177/0032329206290426

World Bank. 1998. *World development report 1998/1999 : Knowledge for development*. New York: Oxford University Press. © World Bank. Available from: https://openknowledge.worldbank.org/handle/10986/5981 License: CC BY 3.0 IGO.

Fanning the Flames or a Troubling Truth? The Politics of Comparison in the Israel-Palestine Conflict

Mandy Turner

ABSTRACT
The politics of comparison in the Israel-Palestine conflict is largely encapsulated in the use of two analogies. The first is the 'Holocaust-Hitler analogy' used by Israel and its supporters, which portrays Israel as a beleaguered nation surrounded by Nazi sympathisers who seek to destroy it as the Jewish homeland. The second is the 'apartheid analogy', which compares the conflict to that of Apartheid-era South Africa and portrays Palestinians as being the victims of racism and settler colonialism. This article analyses why, how and with what desired impact these two comparisons are invoked.

On 30 September 2015, Palestinian Authority President Mahmoud Abbas addressed the UN General Assembly in New York. He warned that Israel was imposing an 'apartheid regime' in the occupied territory – with one legal and governance system for Palestinians, and another for Jewish-Israeli settlers (Times of Israel 2015a). US State Department spokesperson Jen Psaki described the speech as 'provocative' and 'counterproductive', including 'offensive characterizations that were deeply disappointing and which we reject' (Times of Israel and AFP 2015). Unsurprisingly it was widely condemned by Israeli political leaders,[1] with Education Minister Naftali Bennett accusing Abbas of being a Holocaust denier (Pileggi 2015). Less than one month later, on 20 October, Israeli Prime Minister Benjamin Netanyahu, addressing the 37th Zionist Congress in Jerusalem, blamed the Holocaust on the Grand Mufti of Jerusalem, Haj Amin al-Husseini, who was a Palestinian leader during the Second World War. Netanyahu claimed that during a meeting in November 1941, al-Husseini persuaded Hitler to burn Jews rather than expel them from Germany (Prime Minister of Israel's Office 2015). His comments provoked widespread criticism, and motivated German Chancellor Angela Merkel to state: 'We know that responsibility for this crime against humanity is German and very much our own' (Chandler 2015). It also led to a war of words with Palestinian political leaders; Saeb Erekat, former chief negotiator in the peace talks, stated: 'Netanyahu hates Palestinians so much that he is willing to absolve Hitler for the murder of 6 million Jews' (Raved 2015).

These two examples offer a succinct illustration of the focus of this article the politics of comparison in the Israel-Palestine conflict as encapsulated through the use of these two analogies. The first is the 'Holocaust-Hitler analogy' used by Israel and its supporters, which portrays Israel as a beleaguered nation surrounded by anti-Semitic, Nazi sympathisers who seek to destroy it as the Jewish homeland. The second is the 'apartheid analogy', which compares the conflict to that of Apartheid-era South Africa, and portrays Palestinians as being the victims of racism and settler colonialism. The focus of this article is why, how and with what desired impact these two comparisons are frequently invoked. This investigation into the pursuit of political aims through comparing your rival or enemy to evil regimes and monstrous acts is divided into five sections. The first section briefly introduces the use of comparisons in foreign policy decision-making in conflict contexts, then outlines the focus and methodology of critical realism which is the intellectual framework used in this study. The rest of the article applies this framework and methodology to explore how, why, who by, and with what purpose, the Holocaust-Hitler analogy and the apartheid analogy have been used in the context of the Israel-Palestine conflict. The second section, which focuses on the use of the Holocaust-Hitler analogy, unpacks its dual role: first in the development of national cohesion in Israel against an 'evil enemy', and second to build support internationally and to silence critics. The third section, which focuses on the use of the apartheid analogy, explores its dual role: for Palestinians trying to make sense of the social reality they live under; and second to mobilise international support for their emancipation. The chapter concludes with a comparison of these two analogies by returning to the five series of enquiries developed in section one below, i.e. who is invoking each analogy and why, what are the social impacts of their use, to what extent the analogies capture reality, and whether the aim of these analogies is to advance domination or emancipation.

'We're Facing Another Vietnam!' the Case for Analysing 'Truth-Claims' in Conflict Comparisons

There is a body of research on the invocation of comparisons (through the use of analogies and metaphors) in politics, particularly in conflict contexts and especially as relates to foreign policy decisions to intervene in them. During many of the big conflicts of the 20th and 21st centuries, historical comparisons were used by policymakers and opinion-formers (such as journalists) who were sometimes just trying to make sense of events and sometimes using them as an explicit strategy to provoke support and action. While it is difficult to draw a direct link between the use of a comparison and a political decision, some researchers have sought to do so. One of the most comprehensive analyses in this regard is *Analogies at War: Korea, Munich, Dien Bien Phu, and the Vietnam Decisions of 1965* (1992) by Yuen Foong Khong, which looked at the multiple comparisons utilised by US presidents, and how these influenced and guided certain foreign policy decisions. Vietnam, of course, itself went on to become an often-used analogy invoked by US policymakers and opinion-formers, particularly for instance in debates over US policy in Afghanistan (Apple 2001, Miller 2016).

In the post-Cold War world, a recurring comparison in US foreign policy discussions is the Holocaust-Hitler analogy. It has been invoked in two main ways: first, that a stance of non-intervention is effectively abandonment and could lead to genocide; and second, to demonise an opponent or 'enemy'. Secretary of State Madeleine Albright and President Bill Clinton, for instance, used it to rally support for military intervention in Kosovo in 1999 (Akrivoulis 2015, p. 234); while President George W. Bush and Secretary of Defense Donald Rumsfeld invoked it in 2003 to justify regime-change intervention to overthrow the Iraqi Ba'athist regime (Dodge 2016, p. 197–217, Desch 2006, p. 119). In both contexts, as Akrivoulis (2015, p. 237–8) argues, the analogy served three important functions – integration (to rally domestic and international support), legitimation (justify intervention through a powerful moral discourse), and distortion (demonise one side in order to rule out certain policy options, i.e. there could be no negotiated settlement with 'the enemy').

Problematically, however, political scientists often use the terms 'metaphor' and 'analogy' interchangeably. Although both are rhetorical devices in the realm of semiosis (i.e. the making of meaning), there are important differences between the two. The classical understanding of metaphor is that it is a word or concept borrowed from one context and utilised in a different context so that we understand something in terms of a 'transferred meaning' (Maasen and Weingart 2000, p. 19–20). For example, that a person is 'as fit as a fiddle', or that 'life is a rollercoaster'. A man is obviously not a musical instrument (and how can a fiddle be 'fit'?) nor is life a fairground attraction, but the use of such metaphors allow us to 'see things' in a certain way. The use of metaphors also requires shared understandings that are socially, culturally and historically specific, i.e. for them to work, the audience requires prior knowledge and understanding of the metaphor. In the cases referred to above, people would need to know what a 'fiddle' is (and also that this is colloquial for a violin) and that the rollercoaster is a fairground attraction that goes up and down (causing both fear and joy) in order for the comparisons to work.

The two examples given above are benign and harmless, but there are also 'transferred meanings' that are malevolent and harmful. For instance, when politicians or commentators use certain metaphors to build a repertoire of immigrants or refugees as 'unwanted invaders' the implication is that they do not belong in the societies in question and should be driven out or prevented from entering (Parker 2015; Shariatmadari 2015). Metaphors can therefore be effective rhetorical devices employed to create a particular understanding of social reality, and can become a guiding force for future actions (Lakoff and Johnson 1980, p. 75). But metaphors remain suggestive and hypothetical until they are confirmed or corroborated, and so their truth-claims cannot be taken for granted. An analogy, on the other hand, is a logical argument that seeks to demonstrate shared characteristics, and thus it normally contains a 'sense of essence' (Hofstadter 1995, p. 75).

Let us remain with the refugee example for an illustration of the difference between metaphor and analogy. In 2015, UK media commentator Katy Hopkins referred to refugees fleeing war and famine as 'cockroaches'. Obviously, they are not literally cockroaches, but what Hopkins did was employ the *metaphor* of an insect popularly understood as a dirty pest in order to 'transfer meaning' so that the reader would be suitably disgusted and conclude that refugees should be excluded or exterminated. In response, UN High Commissioner for Refugees Zeid Ra'ad Al Hussein then employed an *analogy* to criticise Hopkins' comments by accusing her of using language akin to that of Nazis and Hutu genocidaires (Stone 24 April 2015). While Hopkins is neither a Nazi nor a Hutu genocidaire, her language is certainly reminiscent of them, and so there is a 'sense of essence' to Al Hussein's comparison that cannot be said for Hopkins'.

The debate about whether, and to what extent, there is a 'sense of essence' in any concept and how it relates to social reality is one that deeply divides social and political theorists. Post-structuralists reject the idea of a 'sense of essence' and idealise the 'intersubjective production of meaning' that 'conflates discourses and material practice' (Fairclough *et al.* 2002, p. 2). For post-structuralists, therefore, one analogy or metaphor would have just as much legitimacy as another because there would be no ability to assess whether and to what extent it makes truth-claims and captures reality; in fact, they would reject the pursuit of such an analysis. They would only focus on understanding the way in which the analogy or metaphor was used and for what purpose, and the intentions of those employing it. While these aspects form an important part of the analysis in this article, by rejecting the possibility of assessing truth-claims and 'essence', a poststructuralist perspective restricts the analysis of conflict comparisons in ways that hampers understanding.

Critical realists, on the other hand, insist there are not only 'signifiers' (e.g. words) and 'signified' (concepts) but there are also 'referents' (the material world). Critical realism operates with an ontology that distinguishes between the 'real' (objects, their structures, and their causal powers), the actual (what happens when these powers are activated and produce change), and the empirical (a subset of the 'real' and the 'actual' that is experienced by actors). So while the 'making of meaning' is a central aspect of social reality, it does not constitute it. The 'making of meaning' is 'both socially-structuring and socially-structured', which means that the stories we create to understand various situations structure and 'make' meaning, but that these 'meanings' are *constructed* and only make sense because of a shared history, culture, and social relations (Fairclough 2003, p. 16–30). These will, of course, vary from case to case, thus allowing a range of interpretations. Akin to post-structuralists, critical realists are interested in how these meanings are constructed and the motives of those who employ them. However, in addition and in distinction, they are also interested in the truth-claims and appropriateness of these 'meanings' (Fairclough *et al.* 2002, p. 5).

Because metaphors and analogies are 'effective elements in the interplay of power/knowledge' (Maasen and Weingart 2000, p. 21), we should not just analyse how they emerge and become dominant, but also how societal power relations are either reinforced or challenged by their use, and whether domination or emancipation is the desired object and outcome of their application. For example, referring to refugees as 'parasites' is designed to provoke disgust and violence against a powerless sector of society, while leaving uncriticised and intact what created them in the first place.

This article therefore employs a critical realist perspective in its analysis of the use of metaphors and analogies in conflict comparisons. It adapts aspects of Fairclough's critical discourse analysis (CDA) and utilises insights from Fairclough, Jessop and Sayer's quest to integrate semiosis with critical realism. Fairclough's (1995) CDA framework maps three separate forms (and levels) of analysis on to one another: i) at the micro level – of spoken or written language texts (e.g. through the use of metaphor and rhetorical devices); ii) at the meso level – of processes of text production, distribution and consumption (i.e. who is producing it and who is consuming it); and iii) at the macro level – of instances of sociocultural practice (i.e. the broader environment in which the discourse is being received and perceived, accepted or rejected, acted on or not). Also important to this framework is Fairclough, Jessop and Sayer's insistence on identifying and exploring the 'extra-semiotic conditions' that make semiosis possible and potentially effective, i.e. the influence of the 'habitus' (abstract mental habits and schemes of perception) which is culturally produced; as well as the 'truthfulness' and appropriateness of certain analogies and comparisons, i.e. their relationship with the material and social world (Fairclough *et al.* 2002, p. 5–8). This analysis therefore adopts a verificationist perspective, i.e. that metaphors and analogies should be empirically testable. However, it also recognises that even if the comparison employed does not bear much resemblance to reality, it still has political significance because it creates a framework of understanding and potentially a call to action.

The proceeding analysis of the use of analogies in conflict comparison in the case of Israel-Palestine is guided by five series of enquiries developed from these insights: first, on the way in which the analogy has been invoked and by whom; second, on the intentions of those employing it; third, on the social causes and effects of its use; fourth, on whether and to what extent it captures reality or an 'essence' of the situation to which it is being compared; and fifth, whether the intentions of its invocation are to dominate or to emancipate. The main source of data is literature on the use of these analogies, news articles, speeches, plus visits to Yad Vashem in Jerusalem and Mandela Square in Ramallah. The following section uses these five series of enquiries to explore the use of the Holocaust-Hitler analogy as applied to Palestinians.

The Holocaust-Hitler Analogy and the Politics of 'Never Again'

The Holocaust was the state-sponsored persecution and murder of around six million Jews (as well as other groups such as the Roma, communists, and gay and disabled people) by the German Nazi regime and its collaborators in Europe. It devastated and nearly wiped out European Jewry through mass extermination in gas chambers, after forced displacement into ghettoes, transit camps, and labour camps. The Nuremberg Trials, held between November 1945 and October 1946, prosecuted prominent leaders of the Nazi regime, advanced international law as pertains to war crimes, and were the first forums for the prosecution of crimes against humanity, including the crime of genocide. The Holocaust provoked the creation of the UN Genocide Convention in 1948, which most countries have incorporated into their domestic laws, and which was transferred into the Rome Statute of the International Criminal Court.

It is clear that the Holocaust has had a decisive impact on the identity, ideology and politics of Israel (Segev 2000). Barely a day goes by where there is not a reference to it by Israeli politicians and other opinion-formers such as journalists; and it is often invoked against those regarded as enemies of the Israeli state. As such, only a small selection of examples of its use are utilized here in order to shed light on the five series of enquiries as outlined in section two above; furthermore the focus is on its use against Palestinians.

A visit to Yad Vashem, the Holocaust museum at Mount Herzl on the outskirts of Jerusalem, is an appropriate place to start. 'From Holocaust to Rebirth' is the powerful narrative woven while visitors are led through 10 exhibition halls, ending with the miniature reconstruction of a Nazi gas chamber and photos of the emaciated bodies of those liberated from the camps, next to pictures of David Ben Gurion (first Prime Minister of Israel) making Israel's Declaration of Independence speech on 14 May 1948. Upon exiting the museum, the 'Connecting Path' leads you from Yad Vashem to Har Herzl, which hosts the Mount of Remembrance, i.e. the graves of soldiers killed in combat, victims of terrorist attacks, and Zionist leaders. Along this route are information displays relaying the history of Jewish persecution (most prominently the Holocaust), the emergence and history of Zionism, the Zionist struggle in the final days of the British Mandate, and the birth of the State of Israel – thus linking all of these events in one continuous historical narrative.

The main take-home narrative from Yad Vashem is that the establishment of the State of Israel liberated Jews from millennia of discrimination and genocide. Soldiers in the Israel Defense Forces are given tours by soldier-guides who present a strong narrative that Israel and the Israeli army plays a central role in protecting Jewish lives (Israel Defense Forces 2014). Soldiers also visit Holocaust survivors and Holocaust sites (Efraim 2014; Israel Defense Forces no date). Holocaust education in Israel plays a central role in tying the survival of the Jewish people to the existence of the state of Israel, and in constructing the fear of an existential threat for Jews (Gur-Ze'ev 2000, p. 378). This is passed on through the education system, popular culture and the media. For instance, every year since 1988, 10,000 Israeli high school students, accompanied by Holocaust survivors, visit Yad Vashem and Holocaust sites (Klar *et al.* 2013, p. 128). The Holocaust has therefore played a crucial role in the formation of an Israeli national narrative and sense of loyalty to the state. But, according to critics, this collective historical memory has also fuelled a trauma of victimhood and a type of 'siege mentality' (Magal *et al.* 2016, p. 1224).

In addition to its role in the Israeli national consciousness, the Holocaust has also been invoked as an analogy to demonise opponents and 'enemies'. Comparing Palestinians to one of the most evil regimes of the modern period which explicitly set out to exterminate Jews in Europe has become an important part of Israel's politics of nationhood: it was a powerful mobilizing tool to help promote and justify the violence against Palestinians that made the birth of the state of Israel possible, and it created an ever-present existential threat thus providing a rationale for the continued use of violence to the present day. Zertal (2005, p. 4) argues that: 'There has not been a war in Israel, from 1948 till the present ongoing outburst of violence which began in October 2000 [the book was published in 2005], that has not been perceived, defined, and conceptualized in terms of the Holocaust.' Comparing Palestinians to Hitler is a powerful tool because, as Normand (2016, p. 4) argues, 'he has become the ultimate symbol of evil.' And so Netanyahu's accusation, referred to in the introduction to this article, about al-Husseini being one of the driving forces behind the 'Final Solution', was merely a reiteration of similar accusations made by all previous Israeli leaders right back to Ben Gurion to describe Palestinian leaders and wider Palestinian society. In August 1947, for example, Ben Gurion equated Arabs (including Palestinians) with Nazis and warned that they 'know only one way of solving the Jewish problem: total destruction' (Zertal 2005, p. 173). Menachem Begin, Prime Minister from 1977 to 1983, made prominent use of Holocaust imagery in his speeches, warning time again about the 'return of Auschwitz' in reference to threats from Palestinians and Arab nations (Alexander and Dromi 2011, p. 19). In fact, all Palestinian leaders and many Arab leaders from different eras have been compared to Hitler and/or been accused of being Nazis

(Normand 2016, p. 114). In the run up to the 1967 Arab-Israeli War, Israeli soldiers and politicians frequently claimed that Israel was facing a Holocaust-like disaster. In the context of the bombastic speeches of Egyptian President Gamal Abdul Nasser, this helped to promote worldwide support for Israel which, in this circumstance, was regarded to be battling for its existence against the combined strength of its Arab neighbours.

United States support for Israel was cemented after its success in the 1967 War and grew because of Israel's role as a strategic regional ally (Pappe 2007). From then on, awareness of the Holocaust grew in mainstream American culture to eventually take the dominant position it holds today (Novick 1999). The Holocaust has become an important part of American discourse and politics, including foreign policy decisions, particularly on Israel and Palestine (Finkelstein 2000). In 1994, for instance, former President Richard Nixon wrote: 'No American president will let Israel go down the tubes [because] Israel is a haven for millions whose families endured incredible suffering during the Holocaust' (Desch 2006, p. 108). The Holocaust has thus become a central narrative practice which embeds the idea that the survival of Israel is essential for the survival of the Jewish people, particularly in the collective mind-set of its most powerful ally, the United States. Englert (2018) also charges that this narrative practice allows Western states to sidestep their history of endemic anti-Semitism by giving them a 'quick-fix' through supporting Israel.

When Israel expanded its control over the whole of Mandate Palestine after the 1967 War by occupying the West Bank and Gaza, the comparison was still employed. In 1969, Israel's foreign minister Abba Eban compared a return to the pre-1967 'borders' (the 'Green Line') of Israel with a return to the borders of Auschwitz (Alexander and Dromi 2011, p. 20). Holding on to the OPT has been repeatedly portrayed as being essential for Israel's survival by most Israeli politicians (and the settler movement). This reference to the 1967 'borders' being 'Auschwitz borders' has been asserted on a number of occasions despite the fact that it makes no sense given that Auschwitz had no 'borders' as such. But the imagery created is a powerful one that somehow manages to equate but switch two completely opposite roles, i.e. the colonizer and occupier is transformed into the persecuted victim. For instance, 'One Family', the biggest fund for victims of terror attacks in Israel, had an advertisement on the English website of the Israeli newspaper, *Ha'aretz*, that juxtaposed images of Passover, the Holocaust, and the Second Intifada, which stated: 'In every generation, they rise against us, to annihilate us' (Ochs 2006, p. 360). And to coincide with the International Court of Justice ruling on Israel's Separation Barrier in February 2004, the Israeli Foreign Ministry coordinated a protest outside the court of Israelis who had experienced terror attacks; when a reporter for Israeli Channel 1 asked the widow of a victim for a message, she replied: 'I want to tell the entire world not to close your eyes. Don't ignore this in the same way as they ignored the Holocaust' (Ochs 2006, p. 363).

The main effects of the use of the Holocaust has therefore been to create a sense of existential angst in the Jewish-Israeli population, and has turned the slogan 'never again' into belief in, and support for, a specifically Jewish state with military strength to defeat its enemies. For example, displayed in the office of General Meir Dagan, head of the Mossad from 2002 to 2008, was a photograph of an old Jewish man next to a trench with a rifle aimed at his head by an SS officer. When asked about it, Dagan would reply: 'This old Jew was my grandfather. We should be strong, use our brain, and defend ourselves so that the Holocaust will never be repeated' (Klar *et al.* 2013, p. 135). When this is coupled with the portrayal of Palestinians as akin to the Nazis, an act of collective cognitive dissonance is being employed whereby the heinous act of genocide against 6 million Jews that took place in Europe is transferred to the Arab world in a curious case of the militarisation and geographical displacement of a tragedy. Indeed, Loshitzky (2006, p. 328) argues that: 'The Arab, and particularly the Palestinian, was to become the container of Jewish fantasies of power and revenge.'

Equating Palestinian resistance – both violent and non-violent strategies – with Jewish annihilation is a powerful mobilizing narrative, for both Jewish-Israelis and Jews worldwide. During the Second Intifada, for instance, an Israeli army officer told foreign journalists he was happy to invade Palestinian towns and cities in the West Bank because he was making sure that what happened to his mother, who had been an inmate at Bergen-Belsen concentration camp, would never happen again (Loshitzky 2006, p. 332–3). And in the documentary film 'Israel's Volunteer Soldiers' a Belgian volunteer reveals he was motivated to serve in the Israeli military because he was a grandchild of Holocaust survivors and that 'this was my revenge' (Aljazeera 2017).

There are multiple other examples that could be used here but these suffice for the purposes of this article. It is important for the remainder of this section to assess the extent to which the Holocaust-Hitler analogy captures reality or an 'essence' of the situation to which it is being compared, i.e. whether Palestinians are Nazis and whether Israel is at risk of genocide and annihilation from them.

This article does not seek to assert there were (or are) no Nazi sympathisers or Holocaust deniers in the Middle East, as there clearly were (e.g. al-Husseini) and is (e.g. former Iranian President Mahmoud Ahmadinejad). It is also clear that Nazi symbols (e.g. the swastika) are occasionally used by Palestinian protestors, although less often than you would think, and this is individualised and random, i.e. it is not part of any faction's political platform. The reasons for this are complex: sometimes they are used to identify Israel with the Nazis, sometimes they are used provocatively because it is known that such symbolism is hated by Israelis.[2] But the constant references to Yasser Arafat, Chairman of the PLO, being akin to Hitler do not accord with reality, because it is difficult to substantiate the accusation that he was anti-Semitic. In addition to al-Husseini, the strongest case can be made against current PA President,

Mahmoud Abbas, whose 1982 doctoral dissertation made remarks denying the Holocaust (Achcar 2010, p. 270). But he subsequently acknowledged the tragedy of the Holocaust, stating: '"When I wrote 'The Other Side' [his thesis] we were at war with Israel. Today I would not have made such remarks" (Myre 2003).

The reasons for, and extent of, Holocaust denial in the Arab world have been explored by Litvak and Webman (2009), and Achcar (2010). Litvak and Webman (2009, p. 379) argue that it intensifies during moments of anxiety when Israel's strength is unopposed, whereas eras of diplomacy have brought greater recognition of Jewish experience of persecution and suffering. Achcar criticises the idea that Palestinians bear some responsibility for the Holocaust because they demanded that Britain restrict Jewish migration to Mandate Palestine. He regards this accusation to be hypocritical because other countries, including the United States, also restricted Jewish immigration. Furthermore, he argues that Palestinians should not be blamed for resisting a project whose avowed goal was to establish a foreign state on their land. Achcar (in Berlin, 2015) condemns the use and abuse of the Holocaust-Hitler analogy by Israelis and Arabs alike, but insists that the comparisons drawn between Israel and the Nazis by Arabs are a reaction to decades of Israeli comparisons of Arab leaders with Hitler.

It is difficult not to conclude, therefore, that the main intention behind the use of the Holocaust-Hitler analogy is to create and demonise an 'enemy', to portray yourself as morally superior, and (for some) to justify the use of violence (Normand 2016, p. 114). It is impossible to analyse the international impact of the comparison, but it is clear that most world leaders are sceptical of its use against Palestinians. Nevertheless, its repeated invocation supports the narrative that Israel is the only thing protecting Jews from another Holocaust and it serves to foster guilt amongst Western states at their policies during the Nazi era as well as during and directly after the Second World War, all of which fuels the belief that Israel should be given unequivocal support. As Massad (2006, p. 129) argues: 'Zionism and Israel asserted that any acknowledgment of the holocaust is an acknowledgment of Israel's "right to exist," and conversely that any attempt to deny Israel its alleged right to exist was perforce a denial of the holocaust.' This has helped facilitate the criminalization in the United States and large parts of Europe of criticisms of Israel's oppression of Palestinians. Opposition to the occupation and campaigns for the decolonization of Israel are now presented as constituting a new form of anti-Semitism (Gordon 2018).

But is Israel really at risk of annihilation from Palestinians? Not in the objective sense in that a comparison of military capabilities and casualties leaves no doubt as to who has the upper hand. The Palestinian national movement was a guerrilla organization which was defeated and ejected from Jordan and Lebanon, and which has morphed into a rag-bag of political factions (armed, if at all, with primitive weapons) and Palestinian Authority security forces under the direct control of the US Security Coordinator and tied into security relationships with Israel. Neither PA security forces nor the Palestinian political factions have been able to stop or reverse Israel's colonization of Palestinian land or Israel's oppression of their people. Comparing these actors to Nazis thus strains credulity, but the 'transferred meaning' for Jewish-Israelis when this comparison is made is that they face an existential threat of annihilation. The international message which is transmitted is that Israel faces evil enemies who need to be controlled and/or defeated, and that it should not be criticised for doing so.

In one sense, however, Palestinians *are* a threat against Israel if we remember that Zionism is a movement and state ideology that insists on enshrined legal-political privileges for Jews and regards Jews to be the only people entitled to national rights in the area that used to be Mandate Palestine. However, this is not the same as being a threat to the Jewish people or even to the presence of Jewish-Israelis in a future Palestinian state. In fact, it says more about the Israeli vision of nationhood and its own practices that the two are considered equivalent.[3]

Under current conditions, Israel controls the whole space between the Mediterranean Sea and the Jordan River, but Palestinians in the OPT are living under a separate legal and political regime with no right to vote in Israeli elections and no access to the rule of law. On the other side of the 'Green Line', Palestinians in Israel (who make up 20 per cent of the Israeli population) who lived under military rule until 1966, continue to face discriminatory practices, and are regarded as second class citizens as codified in the July 2018 Nation-State Law (Mendel 2019). Furthermore, millions of Palestinian refugees have been left with no compensation or right of return because Israel refuses to allow it, arguing that their return would destroy Israel as a Jewish state as Jews would thereafter be a minority. While the post 9/11 *zeitgeist* in the United States and Europe has provided a fertile environment for Israel to portray its oppression of a people struggling for liberation and national self-determination as being part of the 'war on terror', critics charge that if Israel does not allow a sovereign Palestinian state to emerge then what will exist is an apartheid regime in which Israel rules over millions of Palestinians who have no rights. The apartheid analogy, which has been invoked by Palestinians and their supporters, is thus explored in the following section.

The Apartheid Analogy and Palestine's 'South Africa Moment'

Apartheid was a system of institutionalised racial segregation and discrimination in South Africa between 1948 and 1991 to ensure white minority dominance over the majority African population. While it only became official state policy in 1948, this system was built on centuries of European settlement and colonialism during which time black South Africans were dispossessed of their land, liberties and rights. It was marked by 'grand apartheid' and 'petty apartheid': grand apartheid restricted movement and residence, which later turned into territorial segregation (the Bantustan 'homelands'), as well as limiting employment opportunities; while petty apartheid restricted interaction between whites and non-whites by segregating public facilities and social events. It was a brutal regime based on discrimination, territorial fragmentation, and political repression. It sparked widespread international condemnation, with the General Assembly of the United Nations designating it a crime against humanity and passing the International Convention on the Suppression and Punishment of the Crime of Apartheid in 1976. In 1977, apartheid was classified as a grave breach of the Geneva Conventions when practiced in situations of armed conflict and as such is a war crime, and it was retained as a crime against humanity in the 1998 Rome Statute of the International Criminal Court (Dugard and Reynolds 2013). The label is more commonly associated in the popular mindset with its experience in South Africa, however, 'apartheid' is actually legally defined. While the analogy has been used in other contexts, it has been more frequently invoked to describe Israel's rule over Palestinians. There are differences in opinion as to its geographical applicability, i.e. whether it should only apply to the Palestinian territory occupied since 1967 (the OPT) which covers Gaza and the West Bank (sometimes including or excluding illegally annexed East Jerusalem), or whether it is applicable to the whole area that covered Mandate Palestine (i.e. Israel plus the OPT including annexed East Jerusalem).

Like the previous section on the Holocaust-Hitler analogy, this analysis of the apartheid analogy, invoked by Palestinians (and some international commentators) to compare their treatment by Israel with South Africa's treatment of non-whites during its apartheid era, focuses on the five sets of enquiries outlined in section two.

An appropriate place to start this analysis is Mandela Square in Ramallah where, in April 2016, a six-metre statue of Nelson Mandela was unveiled. At the inauguration ceremony, South African and Palestinian leaders drew parallels between their liberation movements. Ramallah, the seat of the Palestinian Authority (the non-sovereign governing entity established after the 1993 peace accords), is twinned with Johannesburg whose leaders donated the statue. The Palestine Liberation Organization (PLO) and the African National Congress (ANC) have had a long-standing relationship as anti-colonial movements seeking liberation. In the early 1970s, the PLO and many Palestinian political factions compared Israel's proposals for Palestinian autonomy (expressed through the Allon Plan) to the Bantustan strategy of South Africa (Clarno 2017, p. 3). And leading figures from the South African anti-apartheid movement have continuously drawn parallels between apartheid South Africa and Israel, particularly Nelson Mandela, Archbishop Desmond Tutu, Oliver Tambo, and Ronnie Kasrils. For instance, addressing the UN General Assembly in 1982, Tambo compared the Middle East and southern Africa in terms of the region-wide violence being perpetrated by Israel and South Africa, and the similarities in their liberation struggles (Kasrils 2015, p. 32). In 1997, when apartheid in South Africa had long ended, Mandela said: 'We know too well that our freedom is incomplete without the freedom of the Palestinians' (Prusher 2013). And Tutu (2002, 2012) has repeatedly accused Israel of being an apartheid regime, provoking Alan Dershowitz (2010), US lawyer and prominent Israel supporter, to accuse him of anti-Semitism and of minimising the Holocaust, which indicates how both analogies are frequently invoked as a retort to each other, as also exemplified by the quotations in the introduction to this article. The use of the analogy by prominent South African anti-apartheid activists, however, gives it more credibility and moral weight. Even the architect of apartheid, Hendrik Verwoerd, South Africa's prime minister in the late 1960s, considered Israel to be a kindred spirit practicing an apartheid against Palestinians akin to their policies against black South Africans (McGreal 2006a).

Palestinian and South African leaders have three intentions by invoking the comparison: first, to provide an analysis of their common experiences of oppression; second, to create solidarity and link their anti-colonial struggles; and third, to advance support in international organisations to outlaw such racist modes of governance. In recognition of this, and signalling the activism of developing countries in international affairs in the 1970s, a number of international bodies passed resolutions that compared Zionism with apartheid. In 1975, for instance, the UN General Assembly passed Resolution 3379 which 'determine(d) that Zionism is a form of racism and racial discrimination'; the Organization of African Unity passed Resolution 77 which linked the regimes of Israel, Rhodesia and South Africa to a common racist structure and imperial origin; and the Non-Aligned Movement condemned Zionism as a 'threat to world peace' (Knopf-Newman 2011, p. 41–43).

After 1988 there was less use of the analogy, particularly by the PLO leadership. This reflected changes taking place whereby the PLO gave up its goal of a secular, democratic state in the whole of historic Palestine, instead accepting a solution to the conflict in the creation of a specifically Palestinian state on only 22 per cent of the land. The PLO thereafter *de facto* accepted Israel as a Jewish state even if it has rejected attempts to make it endorse this *de jure*. This change in PLO policy was a precondition of the US-sponsored peace talks between the PLO and Israel. Another Israeli precondition for participating in these talks, was that UN General Assembly Resolution 3379 be rescinded – which happened in December 1991.

But the failure of the peace process, the breakdown of talks at Camp David in 2000, and the growing realisation that Israel was blocking the emergence of a sovereign Palestinian state fuelled the re-emergence of use of the apartheid comparison, mostly from civil society activists within and outside the OPT as well as some of the Palestinian political factions, although not initially from the

Palestinian Authority and PLO leadership. The failure to implement the two-state solution; the huge increase in numbers of Jewish-Israeli settlers in the OPT (who live under Israeli civil law while Palestinians live under military law); and Israel's closure regime of checkpoints, roadblocks and the Separation Wall (referred to by some as the 'Apartheid Wall') has given the comparison renewed vigour and relevance. Indeed, in 2004, Ronnie Kasrils said after visiting the OPT: 'This is much worse than apartheid. The Israeli measures, the brutality, make apartheid look like a picnic. We never had jets attacking our townships. We never had sieges that lasted month after month. We never had tanks destroying houses. We had armoured vehicles and police using small arms to shoot people but not on this scale' (in McGreal 2006b).

In 2001, an attempt was made to resurrect the charge that Zionism is a form of apartheid and colonialism at the World Conference on Racism, which took place in Durban, South Africa. This conference (which was authorised by the UN General Assembly) ended in controversy and is a stark example of the huge divide that exists between Western states and states that experienced Western violence through slavery and colonialism. Efforts to discuss Israel's multiple violations of Palestinian rights were stymied by the withdrawal of the United States and Israel from the conference; many European states also rejected the accusation that Zionism is a form of apartheid.

The year 2005 marks a major turning point in the use of the comparison: the Palestinian Boycott, Divestment and Sanctions (BDS) movement was created and the first Israeli Apartheid Week was established. The BDS movement was launched when 171 civil society groups in the OPT issued a call for boycott, divestment and sanctions against Israel. Drawing inspiration from the anti-apartheid struggle in South Africa and the US civil rights movement, BDS discerned three minimum requirements for a just peace: an end to Israel's military occupation of Gaza and the West Bank (including East Jerusalem) as well as Arab lands in Syria and Lebanon; an end to racial discrimination against Palestinian citizens of Israel; and the right of return for Palestinian refugees as enshrined in UN General Assembly Resolution 194 (Barghouti 2011, p. 49). The BDS movement draws on international law, particularly the 1973 International Convention on the Suppression and Punishment of the Crime of Apartheid and the 2002 Rome Statute. Omar Barghouti (2011, p. 63), one of the movement's leading campaigners, has explicitly called BDS the 'South Africa strategy for Palestine.' Also in 2005, the first Israeli Apartheid Week was held in Toronto, Canada, with a programme of events, speakers and protests that focused on Israel's policies towards Palestinians. By 2019, IAW featured more than 200 events, across 30 countries, on five continents (Israel Apartheid Week website, no date). Using a variety of techniques including installations that resemble the Wall and checkpoints, mock eviction notices, plus lectures and film showings, IAW has been controversial, provoking clashes on university campuses between supporters and critics.

In both cases, the goal of the use of the term 'apartheid' and the comparison with South Africa has been to highlight Israel's practices against Palestinians and mobilise internationally amongst consumers, workers, students, and transnational activist networks. And it is in the realm of civil society that the term 'apartheid' has been invoked most frequently by groups such as B'Tselem, Jewish Voice for Peace, Code Pink, Artists Against Apartheid, and the BDS movement, to name but a few. The stated aims of these groups is to build global support for Palestinian rights and a credible case for a legal challenge against Israel in the International Criminal Court for the crime of apartheid. The role of the apartheid analogy is thus to help institute a 'paradigm shift in the international community's approach to the conflict' (Jacobs and Soske 2015, p. 2). The social effects of the analogy are therefore to be found not so much in the OPT and Israel (although Israel has passed legislation outlawing BDS and is preventing supporters from entering Israel), but more specifically internationally where the battle for public opinion has become increasingly acrimonious.

Just as with the Holocaust-Hitler analogy, journalists have also invoked the apartheid analogy but largely to explore the validity of the comparison. For instance, in 2006, *The Guardian* (London) ran an extensive two-part report written by Chris McGreal (2006a, 2006b) that analysed the charges made by Palestinians and their supporters. McGreal has reported from both Johannesburg and Jerusalem, so was regarded as well placed to make comparisons. McGreal does not offer any firm conclusion as to the legitimacy of the comparison, but there is little doubt that he considers it valid in some ways – through his analysis of Israel's various forms of institutional, legislative and societal discrimination; the unequal education system; land confiscation and unequal access to land; discriminatory planning and urban policy; the differential treatment of Israeli settlers and Palestinians in the OPT; the Wall and policies of separation; the high levels of military violence used by Israel; and Israel's close relationship with apartheid South Africa which continued while the rest of the world was moving towards shunning it. McGreal was criticised for this report, including a complaint to the UK Press Complaints Commission by the pro-Israel lobby group Camera, the Committee for Accuracy in Middle East Reporting in America (which was dismissed). There are many other journalists who use the analogy (e.g. Peter Beinart and Ben White), or reject it (e.g. Melanie Philips and Benjamin Pogrund).

Academics have also invoked the analogy. Early examples include Maxime Rodinson in 1973, and Uri Davis in 1987. But paralleling the recent increase in its use by activist groups, more academic studies have emerged. In 2009, for instance, the Human Sciences Research Council of South Africa (2009), the country's statutory research agency, released a 300-page report concluding that Israel was practicing colonialism and apartheid.

International lawyers have also been more active in the past decade with notable examples coming from two former UN Special Rapporteurs for Human

Rights in the Palestinian Territories Occupied Since 1967: John Dugard (2004–07) and Richard Falk (2008–14). In his 2007 report, Dugard, a South African professor of international law, concluded that some of Israel's practices in the OPT were 'reminiscent' of apartheid South Africa. In a 2016 interview, Dugard explained that he would have used the apartheid label earlier but refrained because he wanted to avoid censure from Western states (Johnston 2007). Falk, professor emeritus of international law at Princeton University, also argued that Israel's regime over the Palestinians 'exhibits features of colonialism and apartheid' (2010, p. 2). Detained at Ben Gurion airport and deported on his first visit to Israel as Special Rapporteur in 2008, Falk faced extensive criticism from Israel and its supporters, including charges of anti-Semitism, despite the fact that Falk is Jewish (Turner 2019).

The apartheid analogy has also been invoked by some unexpected people, for instance by former US President Jimmy Carter in 2006, and by US Secretary of State John Kerry in 2014.

Those who use the apartheid analogy are often accused of being anti-Semitic and of delegitimising Israel (Shimoni 2007). However, this accusation is somewhat blurred by the fact that many Jewish-Israeli politicians and former politicians have also used the analogy. Former Prime Minister Ehud Olmert has raised it repeatedly as a clarion call in favour of ending the occupation and for a two-state solution (McCarthy 2007). Former Prime Minister Ehud Barak has warned of Israel being on a 'slippery slope' towards apartheid (Sommer 2017). And in 2013, Justice Minister Tzipi Livni used the term to argue for the resurrection of peace talks (Times of Israel 2015). The apartheid analogy, in these circumstances, is being used as a warning against the Israeli government, which is working towards annexing large parts of the West Bank – a move that would, in the eyes of critics, ensure perpetual conflict with Palestinians.

As with the Holocaust-Hitler analogy, multiple other examples could be documented here, but these suffice to illustrate who invokes it, for what purpose, and with what intention. It is important now to assess the truth-claims of the apartheid analogy – does it represent an 'essence' of the reality to which it is being compared, i.e. is Israel practicing apartheid against Palestinians?

One of the problems with the analogy is that some who invoke it – on both sides of the debate – get caught up in comparing it with the South African experience rather than understanding it as defined under international law. Article 9 of the Rome Statute states: 'The "crime of apartheid" means inhumane acts of a character similar to those referred to in paragraph 1, committed in the context of an institutionalised regime of systematic oppression and domination by one racial group over any other racial group or groups and committed with the intention of maintaining that regime' (UN 2002). While Israel is not party to two of the key treaties (i.e. the Apartheid Convention and the Rome Statute), the designation 'apartheid' has clear implications under public international law for Israel as well as third parties in terms of the duties of cooperation, non-cooperation and assistance. The significance of the apartheid analogy therefore lies in its designation as an international crime and thus its legal prohibition; indeed, apartheid is illegal under the rules of public international law, whereas occupation is not (Dugard and Reynolds 2013, p. 912–913). The term carries a lot of legal and political purchase, which means that when it is used, particularly in international bodies such as the UN, Israel and its supporters have strongly opposed it thus turning the UN into one of the major diplomatic battle-sites for the Israel-Palestine conflict. Indeed, many more UN bodies and employees beyond the UN Human Rights Council (which hosts the UN Special Rapporteur system) have argued that Israel's repression of Palestinian self-determination and human rights exhibits aspects of apartheid. In 2008, for instance, President of the UN General Assembly Father Miguel D-Escoto Brockman urged that the UN use the term apartheid to describe Israel's policies towards Palestinians. And in March 2012, the UN Committee on the Elimination of Racial Discrimination censured Israel under the rubric of apartheid and segregation as prohibited by Article 3 of the Convention (Dugard and Reynolds 2013, p. 869–70; p. 912). However, like South Africa before it, Israel and its supporters (particularly the United States) regards the UN to be biased against it – and so it frequently criticises the international body and has withdrawn its membership from some UN agencies.

While there will continue to be disputes as regards the extent to which the apartheid analogy captures reality or an 'essence' of the situation to which it is being applied, it is clear that the intentions of those invoking it is to advance Palestinian rights and self-determination – whether that be as a sovereign state existing side-by-side with Israel or whether that be in one democratic state with equal rights for all citizens.

Conclusion: Monstrous Metaphors or Analogies with 'Essence'?

Comparisons made through the use of analogies and metaphors are essential parts of our everyday speech, history, and politics by providing explanations and 'shortcuts' for understanding things through the 'transfer of meaning'. It is therefore hardly surprising they are frequently used in conflict contexts, such as in Israel and Palestine. What is crucial to note, however, particularly in this case, is just how closely tied are the two comparisons analysed in this article. The Jewish-American academic Marjorie N. Feld (2013) once said: 'My own work on American Jews and apartheid suggests just how intense and difficult it is to talk about Israel and apartheid, with the Holocaust and its legacy playing a role in nearly every conversation.' While it is difficult to talk about these comparisons as their use provokes anger from both sides, they are the two main rhetorical frameworks through which both sides see their respective victimhood, and this alone demands we assess them. This conclusion therefore returns to the five series of enquiries used to guide this analysis: first, on the way in which the analogy has been invoked and by whom; second, on the intentions of those employing it; third, on the social causes and effects of its use; fourth, on whether and to what extent it captures reality or an 'essence' of the situation to which it is being compared; and fifth, whether the intentions of its invocation are to dominate or to emancipate.

By using comparisons associated with two monstrous regimes – Nazi Germany and apartheid South Africa – there is little doubt that the desire is to establish and demonise an enemy and build support by using comparisons that many will understand, be suitably disgusted by, and mobilise to fight against. For both comparisons, it is political elites, opinion formers, and civil society activists who use them. In the case of the Holocaust-Hitler comparison, however, this also enjoys sponsorship from the Israeli state. The apartheid comparison, on the other hand, has been mostly driven by civil society actors, initially by leaders of the Palestinian resistance and other national liberation movements, and more recently by civil society organisations and opinion-formers (Palestinian, Israeli, and international); it is only lately that Palestinian political elites have again used it publicly, such as exemplified by Abbas's speech to the UN quoted in the introduction to this article.

The social causes and effects of the use of both comparisons, and their relative 'truth-claims' are a bit more difficult to assess. There is no doubt that Israel and its supporters regard the Holocaust-Hitler comparison as justified and necessary because of the prevalence of anti-Semitism and the fact that Israel resides in a region hostile to its existence. But, as explored in the third section, the Holocaust took place in Europe, not the Middle East. It emerged from centuries of discrimination and persecution of Jewish communities, and it was a specifically *European* phenomenon. This is not to deny that anti-Semitism exists in other parts of the world, because it obviously does. However, Palestinians in the OPT are not in a position to threaten the Israeli state militarily in any meaningful way, despite claims to the contrary. Further, the intent and goal of Palestinian nationalism is neither to eliminate the Jewish people (this is a conflict with Israel not with Jews), nor to install a fascist regime (this is a national liberation movement).[4]

The threat, as identified by Israel itself, comes from a potential change in international public and elite opinion which could lead to pressure for it to withdraw from the OPT and to allow a sovereign Palestinian state to emerge. The current political leadership in Israel has ruled this out and its strategy has been to continue colonising Palestinian land, which will require the perpetuation of the military occupation and suppression of Palestinian resistance to dispossession and oppression. It is difficult not to conclude, therefore, that the Holocaust-Hitler comparison, particularly today when Israel's military dominance is undeniable and international support for its existence is strong, is being used to demonise a people whose land Israel is occupying and confiscating in the pursuit of its goal of gaining control over the whole of Mandate Palestine, to justify these processes, and to silence critics. Domination and not emancipation is thus the purpose and outcome of the Holocaust-Hitler comparison.

Indeed, it might be more accurate to classify the Holocaust-Hitler comparison as a metaphor rather than an analogy, if we refer back to the earlier distinction between the two as rhetorical devices in the 'making of meaning'. A metaphor borrows understandings created in one context to 'bring meaning' to another context, and allows us to 'see' things we would otherwise not see. For Jewish-Israelis (and indeed Jewish communities worldwide), reference to the Holocaust or Hitler is a powerful rhetorical device that invokes a painful and devastating collective historical memory. Its use ensures cohesion and has created a sense of existential crisis because this was the worst historical experience of persecution for the Jewish people. But Palestinians are not to blame for the Holocaust and so in the context of the Israel-Palestine conflict, 'the trauma drama of the Holocaust is a recipe for conflict without end' (Alexander and Dromi 2011, p. 31). Indeed, Avraham Burg (quoted in Alexander and Dromi 2011, p. 10), once Speaker of the Knesset and former chairman of both the World Zionist Organization and the Israeli Jewish Agency, blames 'Holocaust consciousness' for the fact that Israelis are not 'sensitive enough to what happens to others and in many ways are too indifferent to the suffering of others.' Furthermore, there is not a consensus on its use; many Jews, both in Israel and internationally, invoke the Holocaust-Hitler comparison to criticise Israel's actions against Palestinians.

The year 2018 was a significant anniversary year – for both Israel and Palestinians. It marked 70 years since the establishment of the State of Israel and the Nakba, and 50 years of Israel's occupation of the West Bank and Gaza. It is clear that Palestinians and their supporters invoke the comparison with apartheid South Africa as a way to understand their circumstances, connect their experience and struggle with others, and to mobilise internationally for an end to their oppression and dispossession. The social effects of the apartheid comparison are therefore potentially significant. The period of the Oslo peace process (1993–2000) saw a relative hiatus in the use of the apartheid comparison while talks took place and optimism drove beliefs in co-existence. Its resurrection has taken place in an era when peace talks have collapsed and Israel continues to occupy and colonise the (already truncated) geographical space supposed to house a Palestinian state. The re-emergence of the analogy has thus gathered support over the past 15 years as civil society activists and opinion-formers seek to resurrect a movement for Palestinian liberation in the context of an internationally sponsored framework that has only delivered peace to one side.

Even supporters of Israel warn that its rule over Palestinians has 'apartheid-like' features. According to Benjamin Pogrund (2014, p. 133–139), a former deputy editor of the South African newspaper the *Rand Daily Mail* (which was closed down because of its stand against apartheid, and Pogrund was frequently on trial for his views), inside the 'Green Line' Palestinians face discrimination but not apartheid, while inside the OPT Palestinians face colonial practices and tyranny. However, argues Pogrund, while the two-state solution remains the official stance of Israel, Palestine and the international community, then it is not apartheid. But he concludes that if Jewish-Israeli settlement expansion continues and no Palestinian state comes into existence, and if Israel annexes the West Bank and further enforces separation and discrimination, then what will exist thereafter *will* be apartheid (2014, p. 139–152). This conclusion seems no different from what Palestinian activists have said, nor figures such as Carter, Kerry, Dugard and Falk, i.e. that there are *apartheid-like* features of Israel's rule over Palestinians that will soon step over the threshold into a full-blown apartheid system.

What is crucial to note, though, is that the crime of apartheid (like the crime of genocide) has been defined and codified in international law and so strict comparisons with its specific South African experience are not helpful for assessing its legitimacy in the context of the OPT. It is possible, therefore, to conclude that the apartheid comparison should be regarded as a legitimate analogy in that it captures an 'essence' of the reality that Palestinians experience. If international legal opinion eventually agrees with this conclusion and charges Israel with the crime of apartheid, the consequences will be significant. The use of the analogy is consequently an important legal strategy for Palestinians (and their supporters) as they press their claims for self-determination, especially in the absence of other avenues to achieve their goals. The use of such a comparison in this context is therefore highly significant as it has the potential to influence whether Palestinians continue to face dispossession and oppression or whether they gain their emancipation.

Notes

1. Apart from, of course, the political leaders of the Palestinian-Arab community in Israel.
2. I am grateful to Toufic Haddad for this point.
3. I am grateful to Sai Englert for this point.
4. I am grateful to Sai Englert for this point.

Acknowledgments

I would like to thank the editors of this special issue, Bruno Charbonneau and Adam Sandor, for putting it together. I am also extremely grateful to Rimona Afana, Sai Englert, Dan Freeman-Maloy, Toufic Haddad, Nadim Khoury, Michael Pugh, Elian Weizman, and the two anonymous peer reviewers, for incisive comments on an earlier draft. However, all perspectives and errors are my responsibility.

Disclosure statement

No potential conflict of interest was reported by the author.

References

Achcar, G., 2010. *The Arabs and the Holocaust: the Arab-Israeli war of narratives*. London: Saqi Books.

Akrivoulis, D.E., 2015. Metaphors matter: the ideological functions of the Kosovo–holocaust analogy. *Journal of Balkan and near Eastern Studies*, 17 (2), 222–242. doi:10.1080/19448953.2014.993260.

Alexander, J.C. and Dromi, S.M., 2011. Trauma construction and moral restriction. *In*: R. Eyerman, J.C. Alexander, and E. Breese, eds. *Narrating trauma: on the impact of collective suffering*. Boulder, CO: Paradigm Publishers, 107–132.

Aljazeera, 2017. Israel's volunteer soldiers. *AlJazeera*. Available from: http://www.aljazeera.com/programmes/aljazeeraworld/2017/08/israel-volunteer-soldiers-170814073507910.html. Accessed 10 April 2018.

Apple, R.W., 2001. Afghanistan as Vietnam. *New York Times*, 31 October.

Barghouti, O., 2011. *Boycott, disinvestment and sanctions: the global struggle for Palestinian rights*. Chicago: Haymarket.

Berlin, E.B., 2015. Everybody's Holocaust. *Jerusalem Post*, 15 May. Available from: https://www.jpost.com/Jerusalem-Report/Everybodys-Holocaust

Chandler, A., 2015. Germany clarifies: we're responsible for the Holocaust. *The Atlantic*, 21 October. Available from: https://www.theatlantic.com/international/archive/2015/10/holocaust-israel-netanyahu-palestinians-jews/411703/

Clarno, A., 2017. *Neoliberal apartheid: Palestine/Israel and South Africa after 1994*. Chicago: University of Chicago Press.

Davis, U., 1987. *Apartheid Israel: possibilities for the struggle within*. London: Zed Books.

Dershowitz, A.M., 2010. Bishop Tutu is no saint when it comes to Jews, Gatestone institute, 20 December. Available from: https://www.gatestoneinstitute.org/1742/bishop-tutu-is-no-saint-when-it-comes-to-jews. Accessed 12 February 2018.

Desch, M.C., 2006. The myth of abandonment: the use and abuse of the Holocaust analogy. *Security Studies*, 15 (1), 106–145. doi:10.1080/09636410600666295.

Dodge, T., 2006. Enemy images, coercive socio-engineering and civil war in Iraq. *In*: M. Turner and F. Kuehn, eds. *The politics of international intervention: the tyranny of peace*. Abingdon: Routledge, 197–217.

Dugard, J., 2007. Report of the special rapporteur on the situation of human rights in the Palestinian territories occupied since 1967. *UN Human Rights Council, A/HRC/4/17*, 29 January. doi:10.1094/PDIS-91-4-0467B.

Dugard, J. and Reynolds, J., 2013. Apartheid, international law, and the occupied Palestinian territory. *The European Journal of International Law*, 24 (3), 867–913. doi:10.1093/ejil/cht045.

Efraim, O., 2014. A thousand released IDF soldiers celebrate Chanukkah with Holocaust survivors', YNet News. Available from: https://www.ynetnews.com/articles/0,7340,L-4604498,00.html. Accessed 7 July 2018.

Englert, S., 2018. Judaism, Zionism and the Nazi genocide: jewish identify formation in the west between assimilation and rejection. *Historical Materialism*, 26 (2). Available from: http://www.historicalmaterialism.org/articles/judaism-zionism-and-nazi-genocide

Fairclough, N., 1995. *Critical discourse analysis: the critical study of language*. Abingdon: Routledge.

Fairclough, N., 2003. *Analysing discourse: textual analysis for social research*. Abingdon: Routledge.

Fairclough, N., Jessop, B., and Sayer, A., 2002. *Critical realism and semiosis*. Department of Sociology, Lancaster University. Available from: https://pdfs.semanticscholar.org/b7a0/f19699bd6d415df07586c172cb464b84e141.pdf. Accessed 10 April 2018.

Falk, R., 2010. Report of the special rapporteur on the situation of human rights in the Palestinian territories occupied since 1967. *UN Human Rights Council, A/HRC/65/331*, 30 August

Feld, M.N., September 2013. Israeli apartheid: our South Africa moment: review. *Journal of the Society for Contemporary Thought and the Islamicate World*, 27. Available from: https://sctiw.org/sctiw-review/sean-jacobs-and-jon-soske-apartheid-israel-the-politics-of-an-analogy-reviewed-by-marjorie-n-feld/. Accessed 8 June 2018.

Finkelstein, N., 2000. *The Holocaust industry: reflections on the exploitation of Jewish suffering*. London: Verso.

Gordon, N., 4 January 2018. The 'new anti-semitism'. *London Review of Books*, 40 (1), 18. Available from: https://www.lrb.co.uk/v40/n01/neve-gordon/the-new-anti-semitism

Gur-Ze'ev, I., 2000. Defeating the enemy within: exploring the link between Holocaust education and the Arab/Israeli conflict. *Religious Education*, 95 (4), 373–401. doi:10.1080/0034408000950403.

Hofstadter, D., 1995. A review of mental leaps: analogy in creative thought. *Magazine of the American Association for Artificial Intelligence*, 16 (3), Fall, 75–80. doi:10.1609/aimag.v16i3.1154

Human Sciences Research Council of South Africa, 2009. *Occupation, colonialism, apartheid? A re-assessment of Israel's practices in the occupied Palestinian territories under international law*. Cape Town. Available from: http://www.alhaq.org/attachments/article/236/Occupation_Colonialism_Apartheid-FullStudy.pdf.

Israel Defense Forces, 2014. The soldier-guides of the Holocaust Museum Yad Vashem. *IDF Blog*, 29 April. Available from: https://www.idfblog.com/2014/04/29/soldier-guides-holocaust-museum-yad-vashem. Accessed 19 July 2018.

Israel Defense Forces, no date. Tag: witnesses in uniform. *Israel Defense Forces*. Available from: https://www.fidf.org/how-we-help/education/witnesses-uniform. Accessed 6 June 2018.

Israeli Apartheid Week, *Website*, http://apartheidweek.org/. Accessed 14 April 2018.

Jacobs, S. and Soske, J., eds., 2015. *Apartheid Israel: the politics of an analogy*. Chicago: Haymarket.

Johnston, A., 2007. UN envoy hits Israel 'apartheid'. *BBC News*, 23 February. Available from: http://news.bbc.co.uk/2/hi/middle_east/6390755.stm. Accessed 7 April 2018.

Kasrils, R., 2015. Birds of a feather: Israel and apartheid South Africa – colonialism of a special type. *In*: I. Pappe, ed. *Israel and South Africa: the many faces of apartheid*. London: Zed Books, 23–42.

Khong, Y.F., 1992. *Analogies at war: Korea, Munich, Dien Bien Phu, and the Vietnam decisions of 1965*. New Jersey: Princeton University Press.

Klar, Y., Schori-Eya, N., and Klar, Y., 2013. The 'never again' state of Israel: the emergence of the Holocaust as a core feature of Israeli identity and its four incongruent voices. *Journal of Social Issues*, 69 (1), 125–143. doi:10.1111/josi.12007.

Knopf-Newman, M.J., 2011. *The politics of teaching Palestine to Americans: addressing pedagogical strategies*. New York: PalgraveMacmillan.

Lakoff, G. and Johnson, M., 1980. *Metaphors we live by*. London: University of Chicago Press.

Litvak, M. and Webman, E., 2009. *From empathy to denial: Arab response to the Holocaust*. New York: Columbia University Press.

Loshitzky, Y., 2006. Pathologising memory. *Third Text*, 20 (3–4), 327–335. doi:10.1080/09528820600853761.

Maasen, S. and Weingart, P., 2000. *Metaphors and the dynamics of knowledge*. New York: Routledge.

Magal, T., Bar-Tal, D., and Halperin, E., 2016. Why is it so difficult to resolve the Israeli-Palestinian conflict by Jews? a socio-psychological approach. *In*: E. Ben-Rafael, *et al*., eds. Handbook of Israel: the major debates. De Gruyter, 1211–1239.

Massad, J.A., 2006. *The persistence of the Palestinian question: essays on Zionism and the Palestinians*. Abingdon: Routledge.

McCarthy, R., 2007. Israel risks apartheid-like struggle if two-state solution fails, says Olmert. *The Guardian*, 30 November. Available from: https://www.theguardian.com/world/2007/nov/30/israel

McGreal, C. 2006a. Worlds apart. *The Guardian*, 6 February. Available from: https://www.theguardian.com/world/2006/feb/06/southafrica.israel

McGreal, C., 2006b, Brothers in arms: Israel's secret pact with pretoria. *The Guardian*, 7 February. Available from: https://www.theguardian.com/world/2006/feb/07/southafrica.israel.

Mendel, Y., 2019. A new nationalistic political grammar: Jewish-Israeli society 25 years after Oslo. *In*: M. Turner, ed.. *From the river to the sea: Palestine and Israel in the shadow of 'peace'*. Lanham: Rowman and Littlefield, 159–177.

Miller, P.D., 2016. Graveyard of analogies: the use and abuse of history for the war in Afghanistan. *Journal of Strategic Studies*, 39 (3), 446–476. doi:10.1080/01402390.2016.1145588.

Myre, G., 2003. Man in the news: soft-spoken but not afraid to air views, Mahmoud Abbas. *New York Times*, 11 March. Available from: https://www.nytimes.com/2003/

03/11/world/man-in-the-news-soft-spoken-but-not-afraid-to-voice-opinions-mahmoud-abbas.html

Normand, L., 2016. *Demonization in international politics: a barrier to peace in the Israel-Palestine conflict*. New York: PalgraveMacmillan.

Novick, P., 1999. *The Holocaust in American life*. Boston: Houghton Mifflin.

Ochs, J., 2006. The politics of victimhood and its internal exegetes: terror victims in Israel. *History and Anthropology*, 17 (4), 355–368. doi:10.1080/02757200600914102.

Pappe, I., 2007. Clusters of history: US involvement in the Palestine question. *Race & Class*, 48 (3), 1–28. doi:10.1177/0306396807073853.

Parker, S., 2015 'Unwanted invader': the representation of refugees and asylum seekers in the UK and Australian print media. *eSharp*, Issue 23: 'Myth and Nation'. Available from: http://www.gla.ac.uk/media/media_404384_en.pdf. Accessed 12 June 2018.

Pileggi, T., 2015. Netanyahu blasts Abbas's 'deceitful' UN speech. *Times of Israel*, 30 September. Available from: http://www.timesofisrael.com/netanyahu-blasts-abbass-deceitful-un-speech/

Pogrund, B., 2014. *Drawing fire: investigating the accusations of apartheid in Israel*. Lanham: Rowman and Littlefield.

Prime Minister of Israel's Office, 2015. PM Netanyahu's speech at the 37th Zionist congress 20 October. Jerusalem: Government of Israel. Available from: http://www.pmo.gov.il/English/MediaCenter/Speeches/Pages/speechcongress201015.aspx

Prusher, I., 2013. Mandela was critical of occupation, but fully endorsed Israel's right to exist. *Haaretz*, 9 December. Available from: https://www.haaretz.com/blogs/jerusalem-vivendi/.premium-1.562566

Raved, A., 2015. Historians, politicians slam PM's 'distortion of history'. *YNet*, 21 October. Available from: http://www.ynetnews.com/articles/0,7340,L-4714313,00.html. Accessed 10 April 2018.

Rodinson, M., 1973. *Israel: a settler colonial state?* London: Pathfinder Press.

Segev, T., 2000. *The seventh million: the Israelis and the Holocaust*. New York: Henry Holt.

Shariatmadari, D., 2015. Swarms, floods and marauders: the toxic metaphors of the migration. *The Guardian*, 10 August. Available from: https://www.theguardian.com/commentisfree/2015/aug/10/migration-debate-metaphors-swarms-floods-marauders-migrants. Accessed 9 July 2018.

Shimoni, G., 2007. Deconstructing apartheid accusations against Israel. *Jerusalem Center for Public Diplomacy*, 2 September. Available from: http://jcpa.org/article/deconstructing-apartheid-accusations-against-israel/

Sommer, A.K., 2017. Ehud Barak warns: Israel faces 'slippery slope' toward apartheid. *Haaretz*, 21 June. Available from: http://www.haaretz.com/israel-news/1.796949

Times of Israel, 2015a. Full text of Abbas's 2015 address to the UN general assembly. *The Times of Israel*, 30 September. Available from: http://www.timesofisrael.com/full-text-of-abbas-2015-address-to-the-un-general-assembly/.

Times of Israel, 2015b. Livni: we face South Africa-style isolation because of settlements. *Times of Israel*, 18 January. Available from: https://www.timesofisrael.com/livni-we-face-south-africa-style-isolation-because-of-settlements/

Times of Israel and AFP, 2015. US slams Abbas's 'genocide' UN speech as 'offensive'. *Times of Israel*, 27 September. Available from: http://www.timesofisrael.com/us-slams-abbas-un-speech-as-offensive/

Turner, M., 2019. Richard Falk: "citizen-pilgrim" in the role of UN special rapporteur. *Journal of Palestine Studies*, 48(3), 59–78. doi:10.1525/jps.2019.48.3.59.

Tutu, D., 2002. Apartheid in the holy land. *The Guardian*, 29 April. Available from: https://www.theguardian.com/world/2002/apr/29/comment.

Tutu, D., 2012. Justice requires action to stop subjugation of Palestinians. *Tampa Bay Times*, 30 April. Available from: http://www.tampabay.com/opinion/columns/justice-requires-action-to-stop-subjugation-of-palestinians/1227722

United Nations, 2002. *Rome Statute of the International Criminal Court, Part 2, Article 7*. Available from: http://legal.un.org/icc/statute/99_corr/2.htm. Accessed 2 February 2018.

Zertal, I., 2005. *Israel's Holocaust and the politics of nationhood*. Cambridge: Cambridge University Press.

Privileged Sphere of Comparison: Empire, Methods and Conflict Intervention

Bruno Charbonneau

ABSTRACT

Comparison is a method that cannot avoid that its concepts are approximations made for specific audiences or purposes. Deployed to compare armed conflicts, it naturalises the state as the central unit of analysis and the intervention as the external context. This article proposes a strategy that reveals the colonial legacies that form or perform the units of analysis. The possibility of comparing Côte d'Ivoire and Mali appears only when one considers a prior comparison: between Francophone Africa and other 'worlds'. The construction of a French privileged sphere of intervention is inextricably tied to establishing a privileged sphere of comparison.

Comparison is essential as a mode of knowing and being in the world (Sayer 1992). As a research method, it can lead to asking very useful or new questions (Tilly 1984, Moore 1993). The formal art of comparison is construed as requiring the fixing of concepts or units of analysis, in time and in space, in order to facilitate the comparative analysis (Lane 1997). Yet, the world is 'one of countless, overlapping, interacting traditions–so much that, for sociological purposes, it is extremely difficult to decide which units are to be isolated for purposes of comparison' (Gellner 1973, p. 66). One can continuously refine the units of analysis and thus the possibilities of the comparative method, which will depend upon the level of generality of the concepts being deployed. Methodological individualism is an inevitable end to the search for the perfected unit of analysis, but one that cannot avoid that comparative concepts are approximations, ideal-types, made for specific audiences or for specific purposes. In its search for generalisation or universal laws, the formal comparative method has difficulty including context into the measurement statements (MacIntyre 1972). The objective of generalisable and universal knowledge encounters the methodological challenge of the historically contingent event and unit of analysis.

In this article, I seek to address this methodological challenge: the fact that the units of analysis for comparative methods are first and foremost the result of historical relations and processes that change through time and space. As are the objectives of this special issue, I want to do so in the state-centric context of conflict and intervention studies. I argue for a research strategy that emphasises the need to reveal and historicise the relations and processes that form or perform the very units of comparative analysis. I use the cases of the armed conflicts and French military interventions in Côte d'Ivoire and Mali to show the potential of the strategy. I argue that the possibilities of comparing these cases fully appear only when one considers a *prior* comparison: that between the 'world' of Francophone Africa and other 'worlds'. The historical construction of a privileged sphere of intervention, identified as Francophone Africa where France has played and continues to play a pivotal role in conflict management, constitutes the 'time and space relations' that have contributed to the historical production of Côte d'Ivoire and Mali. In short, I show how the historical production of a French privileged sphere of intervention is inextricably tied to the possibilities of comparison within the Francophone Africa 'world' or between 'worlds'. I aim to demonstrate, in other words, both the potential of the research strategy and how methodological choices are inevitably linked to power structures and relations.

The Politics of the Formal Comparative Method

As Heloise Weber (2007) put it, 'a critical evaluation of the formal comparative method is necessary not merely to rectify a methodological problem, but also to expose the politics of methodological choices'. The production of knowledge coming out of the formal comparative method has been, certainly since World War II, inextricably linked to policymaking. In particular, the development of American social and political science led to the creation of 'area studies', of the 'Third World', of 'failed states' and their associated units of analysis that enabled comparisons with the developed world of functioning and democratic states (Seidelman 1985, Ricci 1987). These comparisons constituted the mainstream range of authorised knowledge for representing postcolonial states and societies (Grovogui 1996), representations which in turn authorised the practices of (Cold War) policies and interventions (Bilgin and Morton 2002).

A key critique of the formal comparative method points to what Michael Shapiro (2004, p. xi) termed 'the "cognitive imperialism" of a state-centric social science, which has been involved in attempting to universalize and render self-evident its practices of space and intelligibility.' The formal comparative method naturalises, first and foremost, the state as the space of legitimate political activity and as the central unit of analysis. One practical effect of this approach has been to analyse the postcolonial state in terms of what it should be, in comparison to states in the West, instead of what it is (Hameiri 2010). Methodological nationalism institutes scientific practices that 'bracket out the constitutive character of empire, colonialism and race in global order' (Shilliam in Walker *et al.* 2018, p. 94). Colonial legacies are extracted from political analysis, enabling attempts 'to compare the capabilities of the post-colonial state with the institutional capabilities of states in the West' (Bilgin and Morton 2002, p. 63).

While this state-centric critique is well-known, it is insufficient. The other side of the coin is international intervention. By disarticulating the construction of the postcolonial state from its historical relations, conventional analyses frame conflict, insecurity, poverty, state failure, and so on, 'as originating from "within", disconnected from colonialism and its legacies as well as from the transnational collusions of elite power' (Weber in Walker *et al.* 2018, p. 98). As armed conflicts are posited to originate from the internal issues of postcolonial states, they can be conceived as threats to Western states or to international order. A common assumption is that formal peace processes, formal mandates, and formal international intervention can stabilise armed conflicts and lead to 'peace'. Because the sources of conflict are conceived as 'internal', military interventions are portrayed as 'external' solutions which works to conceal the transnational characteristics and politics of conflict and intervention (Charbonneau 2014). Hence, any analysis of how external military interventions shape and are shaped by the conflict's underlying configurations of power are obscured or dismissed (Cheng *et al.* 2018). International military interventions are not 'external' solutions to 'internal' problems, because they become integral to conflict dynamics as soon as they hit the ground, if not sooner. The naturalisation of the state as a unit of analysis therefore also naturalises 'international intervention' as the external context of conflict management and resolution.

Yet, the spatial limits and assumptions also evolve through, and involve, temporal relations and processes. As the 'terrors of temporal contingency have been tamed by powerful architectures of predictability, often legitimized by various claims that politics must be subordinated to unchanging universalisms' (Walker in Walker *et al.* 2018, p. 91), the challenge is to compare while simultaneously considering the everchanging historical context and contingency of the 'units of analysis'. How can we compare when time and space change and cannot be 'abstracted from the construction of analytical units and categories' (McMichael 1990, p. 395)? According to McMichael (1990, p. 395), the strategy is to subordinate comparison 'to the development of historically-grounded theory rather than using it to establish a causal logic that is generalizable outside time and space relations.' Instead of using fixed units of analysis, one should ground 'the analytical units of comparison in the world-historical processes under investigation' (McMichael 1990, p. 385). Comparing time and place occurrences is a 'substantive historical problem' (McMichael 1990, p. 389) that must reveal and historicise the continuities and legacies of representations, relations, processes, and structures that form or perform the very units of analysis that make comparisons possible.

In what follows, I argue that the possibilities of comparing Côte d'Ivoire and Mali necessitates the analysis of a prior comparison: that made between the 'world' of Francophone Africa and other 'worlds'. The 'world' of Francophone Africa and the states of Côte d'Ivoire and Mali are contingent units of analysis that are embedded in France-Africa power relations. The historical construction of a privileged sphere of intervention, that of Francophone Africa where France has played and continues to play a pivotal and unavoidable role in conflict management, constitutes McMichael's 'world-historical processes' and 'time and space relations' that have contributed to the historical production of the units of analysis under consideration (i.e., the states of Côte d'Ivoire and Mali). In other words, I argue that the (French) privileged sphere of intervention is inextricably tied to the historical production of a privileged sphere of comparison called Francophone Africa which is inherited from French imperialism, and which enables and enacts the tolerable parameters of comparison. This is not a claim about the im/possibility of comparative analysis, but rather one about the politics of such an analysis, about the relationship between theory and practice, and about developing a historically-grounded theory for comparison.

A Privileged Sphere of Comparison/Intervention

Certain typological practices lend themselves well to comparison. 'Francophone Africa' is one, born out of French colonialism and empire. While 'Francophone Africa' is more than the remains of the French empire, it is, at the very least, a reflection of a world once divided between European empires. This space called 'Francophone Africa', which officially designates French ex-colonies (and also includes, at times, Belgian ex-colonies; see Manning 1999), invokes particular colonial legacies and power relations. This space offers the possibility of comparison with other empires, usually with the British empire (see Gaulme 2001, Thomas 2014). While comparisons of diverse imperial practices can be useful to understanding dynamics of conquest, domination, and control, they also carry the risk of producing and reproducing what became known, after decolonisation, as France's *pré carré* (meaning 'enclosed area of influence'). The production of 'Francophone Africa' as a particular 'world' born of empire that can be distinguished from other worlds/empires is intimately tied to the production of a privileged sphere of French influence and intervention, and vice versa. In other words, the act of comparison can work to justify or reproduce an exceptional zone for French military operations.

The construction of 'Francophone Africa' is intimately linked to violent practices that sought, in various ways, to enforce or impose particular dynamics of inclusion and exclusion (Thomas 2011). However, violence could never impose total control and domination or create hermetic 'us' and 'them' categories, because colonial authority worked best when the colonised recognised it or contributed to its implementation (Cooper 2005).

In the context of imperial relations, French and African military practices were closely connected, even if asymmetric and resolutely racist and paternalistic (Clayton 1988, Joly 2009). Decolonisation marked the restructuring of these practices and relationships (Chafer 2001, p. 167). Old capabilities and human relationships found new logics first, in the 'special' Franco-African relationship, and second, in the context of the Cold War. Post-decolonisation Franco-African 'military cooperation' agreements were signed to build the former colony's national armies in order to avoid accusations of neocolonialism and to support the sovereignty claims of the new African states. The transition process from an imperial model of Franco-African troops and armies to a model of national armies was gradual, sometimes tumultuous, was intertwined with the political process of decolonisation, but ultimately failed building functional African armies (Bagayoko-Penone 2003, p. 164–172). Yet, given the very limited resources of African states, the French military quickly became a dependable source of support to consolidate the regimes of African elites who looked favourably upon French involvement in Africa.

France's involvement included some 35 military interventions in Francophone Africa from political independence in 1960 to 1995, usually to sustain or restore to power governments that were friendly to France, a number that excludes dozens of regular military exercises on African soil (Charbonneau 2008). Its military activism earned it the reputation of the 'gendarme of Africa'. The Cold War justified the *pré carré* as it carved it out as a privileged sphere of intervention where France was the guarantor of Western interests in Francophone West and Central Africa. The end of the Cold War meant that France had to adapt to a new normative environment and seek multilateral frameworks of intervention, for example through the EU, the UN and the AU. The dynamics and practices of French military policy in Africa have continuously evolved to the point that by the beginning of the new millennium, the notion of a distinctive francophone space for French military interventions had been considerably watered down (Charbonneau 2017b). Nevertheless, in 2011, the French military ended the Ivorian election crisis and, in 2013, deployed substantial troops in Mali and the Central African Republic. At the time of writing (September 2019), the French military has approximately 5455 troops deployed on operations, 3050 *forces de présence* (permanent troops on French bases in Senegal, Côte d'Ivoire, Gabon and Djibouti), and 1700 'sovereignty' troops at La Réunion, for a total of 10 205 soldiers in Africa.[1]

This continuous French military presence is why all analyses, comparative or not, of armed conflict in Francophone Africa must consider the transnational politics of French involvement. Any appreciation of military interventions in Francophone African conflicts must consider Franco-African postcolonial dynamics because the French military and government are always participants (Powell 2017). Imperial legacies, no matter how defined, are an inevitable influence and the inescapable a priori context of armed conflicts and international interventions in 'Francophone Africa', even if the context is continuously evolving and in constant interaction with non-francophone dynamics and processes. To take into account this context in the francophone space of intervention is to acknowledge the conditions of possibility at play prior to international intervention, how these conditions change according to local/international dynamics, strategies and tactics, and how the limits to peace are set or challenged within this particular context. It is a particular space and moment of inclusion and exclusion, where the distinction between French and African is often blurred by the workings of transnational elite networks (Bancel *et al.* 2003). Yet, it is also a space that can conflate, interact with, or distinguish itself from other global patterns and sites of inclusion and exclusion (Charbonneau and Chafer 2014).

In sum, these are the historical relations, processes, and structures that form and perform the units of analysis – the states – being compared in the next sections: to discuss the Ivorian and Malian states is to ground them in context-specific time and space relations. In other words, their comparison must be grounded *first* in the 'world-historical processes' that produced the possibilities of talking about Francophone Africa. This is one of the imperial legacies involved in all comparisons within this context: it needs the double analysis of the production of independent 'Franco-African states' *and* the production of a privileged sphere of intervention defined as 'Francophone Africa'. This relates apparently separate processes and dynamics as components of a broader historical conjuncture, as components of world-historical processes. Comparisons help deconstruct a particular experience and politics, but they can also reify it. Indeed, if one reifies Francophone Africa, it becomes difficult to go beyond the 'privileged sphere of comparison', to assess its transformation beyond bilateral relations between France and its ex-colonies. It is therefore crucial to situate historically this privileged Franco-African sphere, just as it is crucial to consider the historical (and conflictual) relations that produced the postcolonial state.

Côte d'Ivoire: Global Governance Meets Neocolonial Relations

On 19 September 2002, former soldiers who had been associated with General Robert Guéï's junta in 1999 tried to overthrow Laurent Gbagbo's government. Unable to defeat the Ivorian army in Abidjan, they retreated to Bouaké (and Korhogo) where they would later organise as the *Mouvement Patriotique de la Côte d'Ivoire* (MPCI) under the political leadership of Guillaume Soro. The French military reaction to the attempted coup was without delay. Prepositioned forces (about 600 troops) deployed immediately to protect French, American, and other foreign nationals (Assemblée nationale 2003a). They were reinforced on 22 September when Operation Licorne was officially launched. In

January 2003, French forces were composed of 2,571 soldiers, 14 helicopters, three transport planes, and 172 troops from Senegal. French military strength reached 4,000 troops later that year, and a peak of 5,300 in November 2004.

The French military intervention did not respond to Gbagbo's call for help against foreign aggression,[2] under the 1961 defence accords between the countries, but officially to answer the Ivorian government's call for military support. The intervention almost immediately froze the conflict by establishing a 'zone of trust' separating the north from the south. It prevented more bloodshed, but it also legitimised what was, a priori, a rather limited armed mutiny. Since the French refused to acknowledge Gbagbo's claim of foreign aggression, dividing the country in two enabled the narratives of a North-South confrontation in which the rebels became a legitimate political party to a conflict and its resolution. By splitting the country, the zone of trust produced a political 'North' distinguishable from the political 'South' that was identified, associated, and thus represented by the *Forces Nouvelles*.[3] As Smith (2003, p. 120) puts it, 'As the months went by, the territory conquered by the [MPCI/*Forces Nouvelles*] transformed itself into a source of legitimacy in the eyes of France.'

This North-South narrative did more than split the country in half. It was the key premise, and foundation, for international engagement in the Ivorian crisis. It was the basis of the French-led 2003 Marcoussis Accords; the same Accords that later structured the mandate of the United Nations Operation in Côte d'Ivoire (UNOCI) and informed subsequent peace accords. Within this narrative, the French military intervention established and enabled its own conditions of possibility: the French government could pretend to arbitrate a dispute between distinguishable, legitimate, and indisputable actors and claims. France could affirm to work for 'African stability' (then-Defence Minister Michèle Alliot-Marie, cited in: Assemblée nationale 2004), 'to serve Africans of Côte d'Ivoire' (then-Foreign Affairs Minister Dominique de Villepin, cited in: Assemblée nationale 2003b), and to be an 'impartial force' (Assemblée nationale 2007, p. 30).

It was how things 'always were' in Francophone Africa: France always came, or had to come, to the rescue. African leaders and populations expected France to intervene, for better or worse. Of course, no one in Côte d'Ivoire really believed that France could be impartial. The French official framing of its own involvement was inherently contradictory since it sent the message that political groups could shortcut their way to power by taking up arms. Everyone knew that French economic and political interests and relations in Côte d'Ivoire meant that it was impossible for the French government to stay completely out of the conflict indefinitely. Put another way, a comprehensive analysis of the Ivorian crisis and of the French intervention remains impossible without considering the historical production of French-Ivorian relations (see Charbonneau 2014).

The Historical Relations of a Conflict

As the economic jewel of the French West African empire since the 1930s, Côte d'Ivoire's economic development strategy after independence followed the project of colonial *mise en valeur* (enhancement). The strategy of postcolonial Ivoirian elites was to rely on the growth of the same plantation economy that had been constructed with the coloniser (Chauveau and Dozon 1988), which also relied on labour mobility from, notably, Mali and Burkina Faso.[4] Until the 1980s, the word 'miracle' was employed to describe Côte d'Ivoire and to showcase it as a modernisation theory success story (Dozon 2000).

The country of President Félix Houphouët-Boigny (from 1960 to his death in 1993) had been since independence relatively stable with considerable growth rates. Houphouët-Boigny often boasted that he had opted for capitalist development which compared favourably to the development models of his leftist neighbours Kwame Nkrumah of Ghana, Sékou Touré of Guinea, and Modibo Keita of Mali. He was pro-west, anti-communist, and an ardent capitalist who often sought to foil any pan-African project. As a minister of all French governments between 1956 and 1959, thus during the colonial period, he campaigned against decolonisation, arguing that economic development was the priority and impossible without French help and support (Koné 2003). In the 1950s, however, his influence and strategy were constantly threatened by the rise of communism and nationalism, so he gradually came to favour a transition towards independence that safeguarded his power. His governance strategy was later summarised as 'Houphouetism' (Akindès 2004), meaning a form of state capitalism built upon patrimonial and paternalistic governing practices, and the skilful use of coercion.

The use of coercion relied on, among other things, French guarantees. The regime promoted sustained co-dependent arrangements with French government and economic actors, with the president unapologetically maintaining close ties with French elites. Côte d'Ivoire was the largest recipient of French foreign aid during the 1960s (Bach 1982, p. 113), as French *coopérants* and businessmen were deeply important to Houphouët's statebuilding and economic development projects. The Ivoirian army was kept small and organised around the cooptation of officers, who would often participate in public works but were not trusted by the regime. Houphouët-Boigny preferred the French military, perceived as a dependable source of support and an 'insurance policy' against coup d'état. In short, Houphouët-Boigny was a master of extraversion strategies (Ricard 2017).

Dependency theorists argued that Ivorian economic growth was a mirage, rather than a miracle. In his seminal study of Ivorian capitalism, Samir Amin argued that the Ivorian 'state capitalism' model meant that the state's weight in the economy continuously and very rapidly increased through public investment and an open-door policy of direct foreign investments (especially

French). The result was an economy without a sufficient material base. It was 'growth without development,' largely dependent on foreign capital and managerial expertise. In short, planters did not have to invest, the urban classes were not rich enough to compete with foreign capital, and the country's elites were mainly administrative (see Amin 1973). Following on Amin's steps, Abou Bamba argued that the French role in Ivorian development should not be overstated. Yet, his analysis of American involvement, and of American-French relations, in Ivorian development led him to the same conclusions: Houphouët-Boigny's political project, the foundation of a modern, independent Côte d'Ivoire, was inextricably tied to transnational relations, to the ways in which Ivorian power structures were the product of the historical formation of dependency relationships and transnational elite networks (Bamba 2016).

The fragility and limits of 'Houphouetism' became obvious in the 1980s. Despite state efforts to alleviate the effects of what was called as a 'crisis of development', Houphouët-Boigny's project slowly crumbled as global commodity markets fell, global financial rules turned against him, and structural adjustment policies exacerbated the situation (Losch 2000, Campbell 2002). As the foundations of the post-1960 social contract and political alliances shook and cracked, opposition and protests rose against Houphouët-Boigny's rule. At first, the Ivoirian state fought to alleviate the effects of the crisis and structural adjustments, with the support of France, but the death of Houphouët-Boigny in December 1993 changed the rules of the game. Unhampered by Houphouët-Boigny's resistance, the French government adopted the Balladur doctrine in January 1994, which devalued the franc CFA by 50 per cent and tied all future French aid to the international financial institutions of the Washington Consensus. Côte d'Ivoire was thus opened to the so-called benefits of neoliberal globalisation (Conte 2005).

Henri Konan Bédié succeeded Houphouët-Boigny, believing in the insurance policy of the French government and military to keep him in power. Faced with the rising tensions caused by the Balladur doctrine – caused by the sharp decline in living standards and in the provision of social services, and growing poverty – he chose a divide and conquer strategy, based on an ethnonationalism of *ivoirité*, to maintain his grip on the state. Unable or unwilling to revise the social contract that Houphouët-Boigny had built and sustained, Bédié used the concept of *ivoirité* to identify 'real Ivorians', to explain away the problems of the country by blaming non Ivorians, and to discard his political opponents (ICG 2004, Bah 2010). As social and political tensions continued to rise, and mired in financial scandals, his xenophobic discourse lost him support from key bilateral and multilateral donors.[5] In this context, the army became the agent of political change, overthrowing Bédié in 1999 and setting up the 2000 election of Laurent Gbagbo.

While the Houphouetist model of governance was clearly at stake during the 1990s, it was always embedded in transnational relationships, from global commodity markets to multilateral and bilateral donors, including the former coloniser and its military apparatus. When the civil war officially started in 2002, these relationships did not disappear. They structured international involvement and impacted the possibilities of conflict management and resolution. The conditions under which, and the ways in which, the French government and the 'international community' intervened were shaped by those prior relationships. Such interventions can thus hardly be conceived as 'external' solutions. The formal action of military intervention, however, did set the basic parameters to the legitimate limits of how these relationships could be transformed or, put differently, on how peace could be achieved.

Intervention Politics

Initial French involvement (2002–2004) benefited the rebels to the detriment of Laurent Gbagbo. French military intervention and the follow-up Marcoussis Accords legitimised the armed rebellion and strengthened their position as the Accords established the basis of a power-sharing agreement. Depicted as the 'new' French African policy, military and diplomatic initiatives were reminiscent of the mindset of neocolonial agents, clearly partial, and confident in imposing solutions deemed to be necessary. During the Marcoussis negotiations, President Jacques Chirac threatened to bring Gbagbo in front of the International Criminal Court. Contradicting the assessment of the Chief of Staff of the armies, General Henri Bentegeat, Minister of Foreign Affairs Dominique de Villepin deemed that the military strength of the rebels warranted a power-sharing approach: it was imperative 'to propose to those [MPCI] with the bulk of military power enough attractive responsibilities so that they accept to disarm' (Assemblée nationale 2003b). Back in Abidjan, the Gbagbo government rejected the Marcoussis Accords being imposed by France. In November 2004, rising tensions culminated in a French-Ivorian 'mini war' that saw Chirac ordering the total destruction of the Ivorian air force, French forces firing from tanks and helicopters 'warning and deterrent shots' on protesters, and the death of 9 French, 1 American soldiers, and between 20 and 60 Ivorians (see Amnesty International 2006, D'Ersu 2007). This was the context in which the United Nations Operations in Côte d'Ivoire was authorised in February 2004 (UNSC Resolution 1528).

The dangers of French militarism had been exposed in Rwanda, critics of French African policy had been vocal for decades, and promises of reforms had been made by every president since Charles de Gaulle (Bat 2012). Yet, it was Côte d'Ivoire that ended an era of French military interventionism in Africa. The November 'mini war' had severely weakened French diplomatic credibility and authority, but also birthed a new conflict management approach. The French government could no longer procrastinate over reforms. Côte d'Ivoire became the 'laboratory', as French officials called it, to transform and justify anew, through multilateralization and Europeanisation, French policy in Africa (see Sénat 2006).

Despite several setbacks, none of it abolished the Franco-African privileged sphere of intervention, or the perceived need for French involvement in regional and international conflict management. The years that followed illustrated the challenges to and limits of African unity in leading the Ivorian peace process, even one supported by the UN and UNOCI (Darracq 2011, Vines 2011, Charbonneau 2014). The 2010 presidential election was supposed to end the 'neither war nor peace' situation (McGovern 2011), to move the country towards peace and reconciliation, but instead created a lethal crisis that brought back the French military. Gbagbo lost the election and held on to power, which forced the winner, president-elect Alassane Ouattara, to use force. In the words of Alexandra Novosseloff (2018, p. 18), 'UNOCI forces were caught in the middle, unprepared ... [as i]t tried to remain impartial, but this was untenable, since its role certifying the [election] results and the decisions of the Security Council led it to protect the elected president'. Unable to end the fighting, and as the number of casualties grew, on 3 April 2011, UN Secretary-General Ban Ki-moon wrote to French president Nicolas Sarkozy asking for French military intervention. French operations enabled the arrest of Laurent Gbagbo, and thus ended the fighting.

It helps that France is perceived as a responsible UN Security Council member or as a *mal nécessaire* (necessary evil) in matters of peace and security in Francophone Africa. As UN Under-Secretary-General for Peacekeeping Operations (2000–2008) Guéhenno (2015, p. 93) put it, 'For most members of the [UN Security] council, France "owned" Côte d'Ivoire, and if there was trouble, it was for France to fix.' Even Gbagbo would likely agree. Testifying in front of the French National Assembly in 2007, he insisted upon the two countries' relationships and upon 'the necessity of the French presence in the country and inviting all those who left [after 2004] to come back' (quoted in: Assemblée nationale, France 2007, p. 31).

Mali: Onto Perpetual War

When Nicolas Sarkozy was elected in 2007, he inherited the mess that Chirac had contributed to in Côte d'Ivoire. After the Ivorian election crisis, Sarkozy claimed that France had done its duty, within the parameters of the UN mandate and after the express request of the UN Secretary General. He added that France did not intend to stay and be the police or gendarme of Africa. The Ivorian crisis, at the time, had seemed to signal the end of an era. Hence, in January 2012, when the Malian crisis began with Tuareg rebels of the *Mouvement de libération de l'Azawad* (MNLA), aided by jihadist groups al-

Qaeda in the Islamic Maghreb (AQIM), Ansar Eddine and *Mouvement pour l'unicité et le jihad en Afrique de l'Ouest* (MUJAO), expelled the Malian army from the north, Sarkozy had no interest in intervention, especially during an election campaign. The Malian state's mediocre management of the crisis and the army's humiliation and perceived lack of government support inspired a coup d'état in March, exacerbating the situation as it paralysed the state and further prompted the rebel to conquer the entire northern territory. In June, the MNLA's allies turned on them, hijacking the rebellion. Imposing their rule and control over the conquered territories of the north, they raised fears of jihadist violence and sharia rule.

The election of François Hollande in May 2012 changed the priorities of the French government, as the new president perceived Mali to be an urgent matter threatening both regional and European security (Erforth 2015, p. 101–184). In January 2013, after months of diplomatic negotiations which produced the prudent UN Security Council resolution 2085 (20 December 2012) authorising an African force to be deployed in Mali, the French unilaterally decided to deploy troops (later supported by about 2000 Chadian soldiers). Arguing that Bamako and Malian territorial integrity were in jeopardy as the jihadist groups started moving south, operation Serval was launched to push them back. Its short-term objectives were to restore order, Mali's territorial integrity, and its democratic legitimacy. As such, Serval was to establish the conditions under which the UN could take over, work on a peaceful transition, and address issues of governance and reconciliation. In April, the UN Multidimensional Integrated Stabilisation Mission in Mali (MINUSMA) was created, and the next chapter of international interventions in Mali began.

Indeed, it quickly became apparent that the French military could not leave. Despite Serval's tactical victories, the security situation in Mali never stabilised (Boeke and Schuurman 2015). In 2014, Serval was transformed into the regional Barkhane operation, which covers the G5 Sahel countries (Burkina Faso, Chad, Mali, Mauritania, and Niger). The military logic was justified on the basis that the terrorist threat was regional, Sahelian, and thus demanded a coordinated and regional strategy. Following a similar logic, the G5 Sahel organisation was also created in 2014 to develop regional cooperation for border control and development projects, but the security agenda quickly took the front stage with the creation of a G5 Sahel Joint Force in 2017 (a force that is yet to be fully operational; on the G5, see Desgrais 2018). UN, French, and G5 deployments share the Sahelian space of operations with several other entities: the EU has several security missions (EUTM, EUCAP Sahel Mali, EUCAP Sahel Niger), the US military has deployed between 800 and 1000 special troops and drones in the region, while constructing a base in Agadez, Niger, and the German and the Italian governments have respectively about 900 and 470 soldiers in Agadez. Dozens of competing 'Sahel strategies' inform what is an extensive international engagement comprised of thousands of troops and hundreds of millions of dollars pouring into Mali and surrounding countries.[6]

Yet, despite such military efforts, since 2015 the reports of the UN Secretary-General on the situation in Mali have continuously reached similar conclusions: the security and humanitarian situations are deteriorating, the implementation of the 2015 Peace and Reconciliation Agreement has stalled, violent extremist groups keep extending their territorial reach and area of operations, and the withdrawal of the state's presence over its territory continues, with the situation in the centre of the country being particularly worrisome (see the latest report, UN 2019). A growing academic consensus points to the mounting risks of atrocities, intercommunal violence, human rights violations, the rise of ethnic and intercommunal conflict dynamics, and the disintegration of the social fabric and trust between communities, notably in the central regions of Mopti and Segou, and at the Mali-Niger border (Centre FrancoPaix 2018, Ibrahim and Zapata 2018, Benjaminsen and Ba 2019). In short, unlike in Côte d'Ivoire, there is no end in sight to the conflict or the intervention.

Since Operation Serval, the international narrative has continuously emphasised the terrorist threat, despite several reports, analyses and research articles demonstrating the counterproductive effects of counterterrorist and military solutions (Trotignon 2017, Karlsrud 2018). This obstinacy is justified by claims about the necessity of counterterrorist operations to create the space for peacekeeping, for the peace process, and for the possibility of development (Charbonneau 2017a, 2019). Intervention in Mali is an extensive international engagement that has transformed regional security management and governance. And while there are disagreements over the mechanisms and procedures to use or to put in place – especially between African organisations – there is little to no debate about the priority and necessity of combating terrorism and violent extremism in the Sahel. Mali has become the rationale and justification for transforming regional governance and the multiplication of institutional structures (on the impact on regional security organisations, see Bagayoko 2019).

The Historical Relations of a Conflict

The imperial legacies and the political history of postcolonial Mali are different than Côte d'Ivoire's, and inextricably linked to those of the West African Sahel. The Sahel can be defined as 'an ecozone that marks the transition between the Sahara Desert to the north and the better-watered savanna to the south' (Cooper 2018, p. 2). Today, it is defined politically, comprising the French ex-colonies of Burkina Faso, Chad, Mali, Mauritania, and Niger (i.e., the G5 Sahel). Historically, the Sahel was a space of transregional exchange, human movements, and communities and settlements that were all interdependent and fluid, and thus resistant to the mapping of European empires (Scheele 2012, Nantet 2013). The French were ambitious and imagined the

Sahel as the centre of their West African empire: their plans involved connecting the Senegal River to the Niger Bend and all the way to Lake Chad, and making Timbuktu the link between West African colonies and Algeria (Simonis 2013, Cooper 2018).

Trade relations were a north-south affair, with goods from the West African coast moving to North Africa through the Sahel. The east-west transatlantic slave trade and the extractive demands of the colonial economy weakened these north-south connections, as the French imperial drive transformed the political economy and the basis of social relations. To pay for taxes, Sahelian farmers planted ever greater areas with cash crops, thus contributing to food shortages and even major famines in 1913–1914 and in 1930–1931 (Bonnecase 2011). Colonial rule also drew upon the racial assumptions of slavery to sustain relations of servitude and dominance. In particular, the French racialised the Arab and Berber conflict, romanticised the desert nomads, all leading to reimagining the Saharan and Sudanic 'races' (Klein 1998, Triaud 2014). In the words of Barbara Cooper, this colonial imagination of race relations

> was superimposed upon the pre-existing logic of religion and racial difference that had suffused enslavement in the region. Blackness was associated in French thought with marginal Muslim credentials and nonblackness with relative orthodoxy, if not fanaticism. French dominance entailed protecting the black populations from contamination by the imagined fanaticism of 'Arab' Islam ... [within this colonial logic] Rhetorically and politically, it was difficult to situate the racially ambiguous and highly mobile Tuareg (Cooper 2018, p. 12).

The colonial economy restructured social relations by gradually changing the characteristics of nobility, by encouraging the rise of patriliny, and by racialising differences (Klein 1998, Cooper 2018). These restructured relations would become the basis upon which the Malian postcolonial state was to be built and imagined.

In the mid-20th century, the West African Sahel was the location of radical nationalisms which sought more autonomy from the empire or full independence (Chafer 2002). Yet, in the 1950s and 1960s, decolonisation and state formation also meant 'a crisis of individual and collective political membership' as independence raised 'questions about boundaries, states, and a new international system' and 'about the limits of political communities' (Mann 2015, p. 91). With the birth of the Republic of Mali in September 1960, unlike in Côte d'Ivoire, independence meant breaking away from France, avoiding becoming a French satellite, promoting African unity, and defending a policy of non-alignment. Concretely, the government of Modibo Keita expelled the French army as he ruptured military and defence cooperation relations. The transfer of men and power from an imperial army into a national army was thus much faster in Mali than elsewhere. It also required finding new partners to provide military training and materiel, which came mainly from the Americans, the

British and the Soviets (Joly 2013). Yet, broken military relations did not suspend French-Malian relations. Monetary and financial aid discussions never ceased, as Keita's government could hardly afford being fully independent (Zobel 2013). French development aid always left the door ajar for strategic and military support (Taylor 2019). Military relations indeed warmed up after the 1968 military coup which brought Moussa Traoré to power, leading to a military cooperation agreement signed between the countries in 1985.

In this context, the northern populations of Mali did not fit well within the national state narratives. Adamant to show its authority and independence, and inspired by its own 'strong national army' myth, the Malian state deployed almost half of its army during the first 1963 Tuareg uprising in the Kidal region. As the North was perceived as inhabited by 'Others', reminiscent of the colonial racial administration, the state's show of force set up a security *dispositif* and practices that would be maintained for decades. Much like the French empire had claimed, the Malian state alleged that northern regions and populations required 'special' control and administration. Competition over political authority in the fragmented postcolonial Malian state emphasised status and ethnic differences, pitting north and south (Boilley 1999, Lecocq 2010). In the 2000s, radical Islamic movements entered this competition, further blurring the limits between political communities and questions about the legitimate bases of political authority.

While the 2013 French military intervention brought Mali and the Sahel into the international spotlight, the region has always been perceived as vulnerable, and thus as necessitating interventions. And while Mali was never considered to be part of the Franco-African networks, its dependence upon transnational relations and networks of elite power goes much deeper than Côte d'Ivoire. After World War II, the ascendance of development discourses and practices transformed the Sahel into 'an object of governance, investigation, and intervention', where one could study 'evolving political affinities and novel forms of postcolonial politics, as expressed through humanitarianism and human rights' (Mann 2015, p. 7). After decolonisation, what was once regular movement and travel became crises of transnational migration, exacerbated by perceptions of worrying demographic patterns and ecological degradation. Gregory Mann demonstrated how, as the 'widening gulf between new states and suffering populations' grew, development experts and NGOs came to Mali to fill the gap, but instead widened it. The once imperial subject became an international development beneficiary: 'Another gap was opening, that between state and "government" in its broadest sense' with NGOs increasingly taking over many sectors of state governance (Mann 2015, p. 207). In the same vein, Isaline Bergamaschi demonstrated that international aid and donors contributed to the 2012 crisis by consolidating this form of transnational governance. Contributing to widening gaps, this transnational form of governmental rationality led to 'a state of affairs that provided some impulse and support to putschists and insurgent groups' (Bergamaschi 2014).

Intervention Politics

In April 2013, the UN Security Council sanctioned Serval post-facto when it created MINUSMA, authorising 'French army troops ... to use all necessary means ... to intervene in support of elements of MINUSMA when under imminent and serious threat upon request of the Secretary-General' (Resolution 2100, article 18). Thus was born a 'division of labour': MINUSMA was responsible for the formal political (and peace) processes, while French forces (and their allies) were responsible for military operations against terrorist groups. Underpinning this division of labour was the presumed possibility of differentiating between non-terrorist and terrorist actors, even though the UN recognised the challenge, if not the impossibility, of making the distinction in Mali (UN 2015, 2018). From a legal and normative point of view, the respective doctrines of UN peacekeeping and counterterrorism allow such a division of labour. Because terrorists are said (or assumed) to act outside the legitimate and tolerable limits of political life (Stampnitzky 2013), it can be logical and justified to authorise two parallel military forces in Mali. The legal difficulties associated with distinguishing missions and UN peacekeepers from combat troops operating in the same theatre are thus solved by producing a language of war against terrorism that 'is indistinguishable from a broader strategy of violence legitimization' (Mégret 2002, p. 364). And as these legal issues find their solution – albeit disputed – the division of labour between the two military forces in turn enables and enacts the political distinction between terrorist and other actors (see Charbonneau 2017a).

As the security situation gradually worsened after 2014, Mali quickly offered the rationale for a restructuring of regional security governance. First, it justified and enabled the latest restructuring of the French military posture in Africa. For the French, the transnational dimension of the Sahel 'war on terror' required a regional approach. Serval was terminated (declared a success) and its forces were integrated with those of operation Épervier in Chad.[7] Operation Barkhane was thus created, spanning the countries of the G5 Sahel. Under Barkhane, French forces have operated 'autonomously' in Mali and, 'in cases of emergency', in Niger. Offensive operations in Burkina Faso, Chad, and Mauritania must first be authorised by their respective governments. Sensitive operations in Algeria, in Libya or against Boko Haram, technically outside the area of operation of Barkhane, can be authorised by Paris. In short, Paris has transformed the old bilateral arrangements with African countries into a unique regional military apparatus that is on a permanent war footing (Charbonneau 2019). Indeed, the French military argues that it cannot totally eradicate 'terrorism', but will continue 'mowing the lawn' for the next '15–20 years', a powerful image for regularly cutting heads (Bigot 2015, Charbonneau 2017b).

The year 2014 also saw the creation of a new African regional organisation: the G5 Sahel. The organisation is the embodiment of the so-called 'security-development nexus' as its purpose is to coordinate international and regional development and security policies (Desgrais 2018). The nexus has been particularly disposed to military logic as it serves the argument that military interventions in the Sahel are necessary to produce the space and time for peace and development. Indeed, and despite numerous discourses of good intentions, the G5 Sahel and its international partners rapidly focused their efforts on the deployment of the G5 Joint Force. A symbol of the emergence of a new regional security architecture (Diallo 2018), and supposed, in some distant future, to replace French troops, the Joint Force is nevertheless dependent on Barkhane's planning, logistics, communication, and strategic transport capabilities.

In a sense, French Operation Serval opened the newest chapter of a long history of international intervention in Malian and Sahelian affairs. The key difference in 2012 was the rise of extremist jihadist armed groups, which justified both the central role of the French military and that matters of political rule and governance in Mali be subjected to military logic. The use of military force has been normalised in Mali and the G5 Sahel countries, as the continuous (military) management and suppression of terrorism and 'violent extremism' are deemed to be necessary measures to building peace and development.

Comparative Analysis

At this point, the typical comparative analysis would point to causal mechanisms or create a matrix to identify similarities and differences between Côte d'Ivoire and Mali. It might look something like this: both countries share a colonial heritage, a common currency, and various political and economic relationships (via la Francophonie in particular). Both conflicts started from armed challenges to state authority and legitimacy, which evolved into political forces who became involved in formal peace processes. Both countries experienced a French military intervention, which led to a parallel UN peacekeeping deployment. In term of differences, the motives of the insurgents differed: rebels in Côte d'Ivoire tried to take over the state, reached a power-sharing agreement in 2007, and came to power in 2011. In Mali, the insurgents were irredentists, with a jihadist twist, posing an existential challenge than one for greater control of the state apparatus. One conflict is understood to be a civil war, while the other is debated as partly a civil conflict, partly as another site of the 'global war on terror'. A basic difference between Côte d'Ivoire and Mali was in the relative strengths of state and non-state formations, which partly explains the different dynamics of transnational relations of elite power. The story in Côte d'Ivoire is some version of liberal peacebuilding extraversion which allows Ivorian transnational elites to consolidate their power and wealth through the capture of various markets and rents (Ricard 2017). The story in Mali is some version of extraversion à la Mann (2015), which the post-Serval interventions have fostered into a quasi-permanent condition of governance dependency that suits the national and international elites.

Such a formal comparative analysis is useful for explaining each case, notably for those interested in 'lessons learned' and policy prescriptions. Yet, it is a tricky proposition because comparing the two conflicts in such a manner involves accepting (as an 'independent variable') the recurring fact of French involvement in African conflicts. It cannot explain, and must overlook, the constitutive relationships of France, Côte d'Ivoire and Mali, and how they establish peculiar processes of international intervention. More interesting and productive, I argue, was my focus on the evolution of these relationships and 'world-historical processes', that is, on the transformation of transnational dynamics that constitute the units of analysis called Côte d'Ivoire and Mali. The comparison becomes something else entirely because it enables the analysis of how the transnational dynamics of one conflict move or evolve into the other, how intervention and conflict politics overlap, how one intervention/conflict informs the other, and how the practices of intervention changed in what is a privileged sphere of (French) intervention/comparison.

Such an approach can show why it is always the same actors, acting in similar fashion, that keep coming back. In 2013, as the French engaged more troops in Mali and Central African Republic, arguably demonstrating once again that the French military remained a relevant if not an inescapable characteristic of Francophone African conflicts, Ivorian president Alassane Ouattara was already hard at work 'normalizing' the situation in the country (Ricard 2017). UNOCI began its drawdown and closed relatively quickly in June 2017. In January 2015, French Operation Licorne was transformed into the *Forces Françaises en Côte d'Ivoire* (FFCI) after Ouattara convinced the French government to keep a military presence in the country. A *force de présence* contingent of approximately 400 soldiers, the FFCI are tasked chiefly with training Ivorian forces and with the logistical support of French Operation Barkhane in the Sahel. Resolute to restore the confidence of foreign investors and sustain economic growth, Ouattara (re)asserted Ivorian diplomatic weight in matters of international security and regional counterterrorism: Côte d'Ivoire obtained a non-permanent seat on the UN Security Council (2018–2020), sent 163 soldiers and 21 police officers to the UN peacekeeping mission in Mali, built with France a counterterrorist regional academy near Abidjan (*Académie internationale de lutte contre le terrorisme*; see RFI 2018), and signed a judiciary accord on counterterrorist cooperation with Mali (Mieu 2018).

Such rapprochement in counterterrorist matters between France, Côte d'Ivoire and Mali demonstrates how the transformation of the regional security architecture produces effects beyond the G5 Sahel zone. It also shows how the African francophone state (whether Côte d'Ivoire or Mali) is constructed as a space of privileged intervention politics, as a space where a specific intervener has special access and understanding, and where it can make judgements that others cannot. In other words, it demonstrates how recurrent French meddling in African conflicts (1960–2011) has turned into a permanent war posture in the West African Sahel since 2013. The construction of the African francophone state cannot be understood outside of the construction of this privileged sphere of intervention/comparison. Within this sphere, the theorisation of and the tolerable limits to violence are not state monopolies but constituted transnationally and sustained by evolving time and space relations. This is a transnational sphere of asymmetric relationships between political, economic and military elites that struggle, compete or cooperate for diverging interests and benefits. In this context, the comparative method is still valuable, not so much as an ahistorical comparison of two conflicts seeking 'causal mechanisms', but instead as a method to show how a privileged sphere of intervention/comparison is sustained through time and space.

Conclusion

My analysis sought to demonstrate how the way we can compare 'units of analysis' has inevitable political effects. Comparing comes at a cost that should not be obscured, hidden, or brushed aside as methodological rigour or choices. That cost must be acknowledged, and its politics revealed.

In this article, my research strategy was to first expose an assumption or a prior comparison point: the Francophone African world. This beginning bestowed and recognised it as a privileged sphere, as something that can be distinguished from other worlds. The analysis showed how Côte d'Ivoire and Mali are reproduced within the relations constitutive of that world, partly based on the historical construction of time and space relations and processes that enable, enact and sustain a particular form of military intervention.

Every beginning cuts and discriminates. Is it possible to do otherwise? My analysis sought to, at least, alleviate the costs and effects of other discriminations found in the formal comparative method: between internal and external, between conflict and intervention, between national and international; and to reveal their politics and multiple effects on our understanding of conflict/intervention. Comparing can certainly reproduce categories, which can then be instrumentalized and deployed both within academia and the policy world. Comparing can also, however, destabilise by exposing how rigid categories can be resources of power. My analysis sought to destabilise what Shapiro named the '"cognitive imperialism" of a state-centric social science' that imposes limits on political possibilities and imagination. It sought to expose how context-specific power structures and relations were reproduced and perpetuated through imperial legacies, transnational relations, and global configurations.

Notes

1. https://www.defense.gouv.fr/operations/rubriques_complementaires/carte-des-operations-et-missions-militaires .
2. The Gbagbo government claimed that the coup was a foreign aggression because the coup organisers had planned their operations in Ouagadougou, the capital of Burkina Faso.
3. *Forces nouvelles* is the named given to the military wing or group associated to the political organisation of the MPCI.
4. Thanks to JS for this.
5. The context of the mid- and late 1990s was that of the aftermath of the Rwandan genocide, the conflicts in Somalia and Bosnia (notably the Srebrenica massacre/genocide), and the 1999 NATO intervention in Kosovo, and thus of the fear of a repeat scenario in Côte d'Ivoire.
6. MINUSMA is authorised to deploy 13,289 military and 1,920 police personnel but has an effective force of 13,155 and 1,738 as of May 2019. Barkhane has approximately 4,500 soldiers based and operating across the G5 countries. The G5 Sahel organisation was supposed to deploy a joint force of 5,000 throughout 2018, which it proved incapable of doing.
7. Epervier was created in 1986 to support Hissène Habré's regime against Libyan forces. Since 1990, French forces have repeatedly intervened to protect the regime of president Idriss Déby – most recently on 3 February 2019 against the Chadian rebel movement, the *Union des forces de la résistance*.

Acknowledgments

I want to thank Geneviève Parent, Maxime Ricard, Adam Sandor, Jonathan Sears, and Mandy Turner for their constructive and encouraging comments on improving this article. I want to thank all the contributors to our October 2017 Centre FrancoPaix workshop on 'Comparing Armed Conflicts' for a most invigorating discussion, which has proved very fruitful. I also want to recognize the support of the Social Science and Humanities Research Council of Canada and the Department of National Defence (Canada) for providing the necessary funding for the organization of our workshop.

Disclosure statement

No potential conflict of interest was reported by the author.

Funding

This work was supported by the Social Sciences and Humanities Research Council of Canada.

ORCID

Bruno Charbonneau http://orcid.org/0000-0003-0373-7174

References

Akindès, F., 2004. *Les Racines de la crise militaro-politique en Côte d'Ivoire*. Dakar: CODESRIA.

Amin, S., 1973. *Le développement du capitalisme en Côte d'Ivoire*. 2nd ed. Paris: Les éditions de Minuit.

Amnesty International, 2006. *Côte d'Ivoire. Clashes between peacekeeping forces and civilians: lessons for the future*. Online.

Assemblée nationale, France, 2003a. *Commission de la Défense nationale et des Forces armées*, 21 Jan, Compte rendu No 26. Paris.

Assemblée nationale, France, 2003b. *Commission des Affaires étrangères*, 29 Jan, Compte rendu no 30. Paris.

Assemblée nationale, France, 2004. *Commission des Affaires étrangères*, 9 Nov, Compte rendu No 10. Paris.

Assemblée nationale, France, 2007. *Rapport d'information par la Commission des Affaires étrangères en conclusion des travaux d'une mission d'information constituée le 13 décembre 2006 sur la situation des Français rapatriés de Côte d'Ivoire*, 13 Feb. Paris: No 3694. doi:10.1094/PDIS-91-4-0467B.

Bach, D., 1982. L'insertion ivoirienne dans les rapports internationaux. *In*: A. Fauré and J.-F. Médard, eds. *État et bourgeoisie en Côte d'Ivoire*. Paris: Karthala, 89–121.

Bagayoko, N., 2019. *Le multilatéralisme sécuritaire africain à l'épreuve de la crise sahélienne*. Montreal: Centre FrancoPaix de la Chaire Raoul-Dandurand.

Bagayoko-Penone, N., 2003. *Afrique: les stratégies française et américaine*. Paris: l'Harmattan.

Bah, A.B., 2010. Democracy and civil war: citizenship and peacemaking in Côte d'Ivoire. *African Affairs*, 109 (437), 597–615. doi:10.1093/afraf/adq046.

Bamba, A., 2016. *African miracle, African mirage: transnational politics and the paradox of modernization in ivory coast*. Athens: Ohio University Press.

Bancel, N., Blanchard, P., and Vergès, F., 2003. *La République coloniale. Essai sur une utopie*. Paris: Albin Michel.

Bat, J.P., 2012. *Le syndrome Foccart. La politique française en Afrique, de 1959 à nos jours*. Paris: Gallimard.

Benjaminsen, T. and Ba, B., 2019. Why do pastoralists in mali join jihadist groups? A political ecological explanation. *Journal of Peasant Studies*, 46 (1), 1–20. doi:10.1080/03066150.2018.1474457.

Bergamaschi, I., 2014. The fall of a donor darling: the role of aid in mali's crisis. *The Journal of Modern African Studies*, 52 (3), 347–378. doi:10.1017/S0022278X14000251.

Bigot, L., 2015. L'opération 'Barkhane', un 'permis de tuer au Sahel'. *Le Monde Afrique*, 2 Nov.

Bilgin, P. and Morton, A.D., 2002. Historicizing representations of 'Failed States': beyond the Cold War annexation of the social science? *Third World Quarterly*, 23 (1), 55–80. doi:10.1080/01436590220108172.

Boeke, S. and Schuurman, B., 2015. Operation 'Serval': a strategic analysis of the French intervention in Mali, 2013–2014. *The Journal of Strategic Studies*, 38 (6), 801–825. doi:10.1080/01402390.2015.1045494.

Boilley, P., 1999. *Les Touaregs Kel Adagh. Dépendances et révoltes: du Soudan français au Mali contemporain*. Paris: Karthala.

Bonnecase, V., 2011. *La pauvreté au Sahel du savoir colonial à la mesure internationale*. Paris: Karthala.

Campbell, B., 2002. Political dimensions of the adjustment experience of Côte d'Ivoire. *In*: A. Bakan and E. MacDonald, eds. *Critical political studies: debates and dialogues from the left*. Montreal: McGill-Queen's University Press, 156–178.

Centre FrancoPaix, 2018. *Stabilizing Mali: the challenges to conflict resolution*. Montreal: Chaire Raoul-Dandurand.

Chafer, T., 2001. French African policy in historical perspective. *Journal of Contemporary African Studies*, 19 (2), 165–182. doi:10.1080/02589000120066443.

Chafer, T., 2002. *The end of empire in French West Africa*. Oxford: Berg.

Charbonneau, B., 2008. *France and the new imperialism: security policy in sub-Saharan Africa*. Aldershot: Ashgate.

Charbonneau, B., 2014. The imperial legacy of international peacebuilding: the case of Francophone Africa. *Review of International Studies*, 40 (3), 607–630. doi:10.1017/S0260210513000491.

Charbonneau, B., 2017a. Intervention in Mali: building peace between peacekeeping and counterterrorism. *Journal of Contemporary African Studies*, 35 (4), 415–431. doi:10.1080/02589001.2017.1363383.

Charbonneau, B., 2017b. De Serval à Barkhane: les problèmes de la guerre contre le terrorisme au Sahel. *Les Temps modernes* 693–694(avril-juillet), 322–340. doi:10.3917/ltm.693.0322.

Charbonneau, B., 2019. Faire la paix au Mali: les limites de l'acharnement contre-terroriste. *Revue Canadienne des études africaines*, 53 (3). Forthcoming.

Charbonneau, B. and Chafer, T., eds., 2014. *Peace operations in the francophone world: global governance meets post-colonialism*. New York: Routledge.

Chauveau, J.P. and Dozon, J.P., 1988. Ethnies et État en Côte d'Ivoire. *Revue française de science politique*, 38 (5), 732–747. doi:10.3406/rfsp.1988.411165.

Cheng, C., Goodhand, J., and Meehan, P., 2018. *Synthesis paper: securing and sustaining elite bargains that reduce violent conflict*. London: UK Stabilisation Unit.

Clayton, A., 1988. *France, soldiers, and Africa*. London: Brassey's.

Conte, B., 2005. La responsabilité du FMI et de la Banque mondiale dans le conflit en Côte d'Ivoire. *Études Internationales*, 36 (2), 219–228. doi:10.7202/011416ar.

Cooper, B., 2018. The Sahel in West African history. *Oxford Research Encyclopedia of African History*. doi:10.1093/acrefore/9780190277734.013.167.

Cooper, F., 2005. *Colonialism in question: theory, knowledge, history*. Oakland: University of California Press.

D'Ersu, L., 2007. La crise ivoirienne: une intrigue franco-française. *Politique africaine*, 105 (March), 85–104. doi:10.3917/polaf.105.0085.

Darracq, V., 2011. Jeux de puissance en Afrique: le Nigeria et l'Afrique du Sud face à la crise ivoirienne. *Politique étrangère*, 2 (summer), 361–374. doi:10.3917/pe.112.0361.

Desgrais, N., 2018. La Force conjointe du G5 Sahel ou l'émergence d'une architecture de défense collective propre au Sahel. *Les Champs de Mars*, 1, 211–220. doi:10.3917/lcdm.030.0211.

Diallo, O., 2018. L'architecture africaine de paix et de sécurité dans le Sahel: entre adaptation et compétition face aux menaces régionales. *Bulletin FrancoPaix*, 3 (5–6), 1–8.

Dozon, J.-P., 2000. La Côte d'Ivoire entre démocratie, nationalisme et ethnonationalisme. *Politique africaine*, 78 (2), 45–62. doi:10.3917/polaf.078.0045.

Erforth, B., 2015. *Thinking security: a reflectivist approach to france's security policy-making in sub-Saharan Africa*. PhD thesis. University of Trento, Italy.

Gaulme, F., 2001. *Intervenir en Afrique? Le dilemme franco-britannique*. Paris: IFRI (Institut français des relations internationales), Notes de l'Ifri 34.

Gellner, E., 1973. *Causes and meaning in the social sciences*. London: Routledge.

Grovogui, S., 1996. *Sovereigns, quasi sovereigns, and africans: race and self-determination in international law*. Minneapolis: University of Minnesota Press.

Guéhenno, J.-M., 2015. *The fog of peace*. Washington: Brookings Institution Press.

Hameiri, S., 2010. *Regulating statehood: statebuilding and the transformation of the global order*. Basingstoke: Palgrave.

Ibrahim, I.Y. and Zapata, M., 2018. *Regions at risk: preventing mass atrocities in Mali*. Washington: United States Holocaust Memorial Museum.

International Crisis Group, 2004. *Côte d'Ivoire: no peace in sight*. Dakar/Brussels: Africa report no 82.

Joly, V., 2009. *Guerres d'Afrique. 130 ans de guerres coloniales, l'expérience française*. Rennes: Presses universitaires de Rennes.

Joly, V., 2013. The French Army and Malian independence (1956–1961). *In*: T. Chafer and A. Keese, eds. *Francophone Africa at fifty*. Manchester: Manchester University Press, 75–89.

Karlsrud, J., 2018. *The UN at war: peace operations in a new era*. Middletown: Palgrave.

Klein, M., 1998. *Slavery and colonial rule in French West Africa*. Cambridge: Cambridge University Press.

Koné, A., 2003. *Houphouët-Boigny et la crise ivoirienne*. Paris: Karthala.

Lane, R., 1997. *The art of comparative politics*. Boston: Allyn and Bacon.

Lecocq, B., 2010. *Disputed desert: decolonisation, competing nationalisms and tuareg rebellions in Mali*. Leiden: Brill.

Losch, B., 2000. Coup de cacao en Côte d'Ivoire, Économie politique d'une crise structurelle. *Critique internationale*, 9, 6–14. doi:10.3406/criti.2000.1614.

MacIntyre, A., 1972. Is a science of comparative politics possible? *In*: P. Laslett, W. G. Runciman, and Q. Skinner, eds. *Philosophy, politics and society*. Oxford: Basil Blackwell, 8–26.

Mann, G., 2015. *From empires to NGOs in the West African Sahel*. New York: Cambridge University Press.

Manning, P., 1999. *Francophone Sub-Saharan Africa 1880–1995*. Cambridge: Cambridge University Press.

McGovern, M., 2011. *Making war in Côte d'Ivoire*. Chicago: University of Chicago Press.

McMichael, P., 1990. Incorporating comparison within a world-historical perspective: an alternative comparative method. *American Sociological Review*, 55 (3), 385–397. doi:10.2307/2095763.

Mégret, F., 2002. 'War'? Legal semantics and the move to violence. *European Journal of International Law*, 13 (2), 361–399. doi:10.1093/ejil/13.2.361.

Mieu, B., 2018. La Côte d'Ivoire et le Mali musclent leur coopération judiciaire pour lutter contre le terrorisme. *Jeune Afrique*, 11 May.

Moore, B., 1993. *Social origins of dictatorship and democracy*. Boston: Beacon Press.
Nantet, B., 2013. *Histoire du Sahara*. Paris: Tallandier.
Novosseloff, A., 2018. *The many lives of a peacekeeping mission: the UN operation in Côte d'Ivoire*. New York: International Peace Institute.
Powell, N., 2017. Battling instability? The recurring logic of French military interventions in Africa. *African Security*, 10 (1), 47–72. doi:10.1080/19392206.2016.1270141.
RFI, 2018. Lutte contre le terrorisme: colloque de l'Académie internationale à Abidjan. *RFI Afrique*, 20 Nov. Available from: http://www.rfi.fr/afrique/20181120-lutte-terrorisme-colloque-academie-internationale-abidjan-bakayoko [Accessed 3 September 2019].
Ricard, M., 2017. Historicity of extraversion in Côte d'Ivoire and the 'post-conflict'. *Journal of Contemporary African Studies*, 35 (4), 506–524. doi:10.1080/02589001.2017.1361511.
Ricci, D., 1987. *The tragedy of political science: politics, scholarship, and democracy*. New Haven: Yale University Press.
Sayer, A., 1992. *Method in social science: a realist approach*. London: Routledge.
Scheele, J., 2012. *Smugglers and saints of the Sahara*. New York: Cambridge University Press.
Seidelman, R., 1985. *Disenchanted realists: political science and the American crisis*. New York: SUNY.
Sénat, France, 2006. *Rapport d'information au nom de la commission des Affaires étrangères, de la défense et des forces armées sur la gestion des crises en Afrique subsaharienne*, 3 July. Paris: No 450.
Shapiro, M., 2004. *Methods and nations*. London: Routledge.
Simonis, F., 2013. Du haut-fleuve au Mali indépendent. *In*: Gonin, Kotlok, and Montclos, eds. *La tragédie malienne*. Paris: Vendémiaire, 37–60.
Smith, S., 2003. La politique d'engagement de la France à l'épreuve de la Côte d'Ivoire. *Politique africaine*, 89 (March), 112–126. doi:10.3917/polaf.089.0112.
Stampnitzky, L., 2013. *Disciplining terror: how experts invented 'terrorism'*. Cambridge: Cambridge University Press.
Taylor, I., 2019. France à fric: the CFA zone in Africa and neocolonialism. *Third World Quarterly*, 40, 1064–1088. doi:10.1080/01436597.2019.1585183.
Thomas, M., ed., 2011. *The French colonial mind: violence, military encounters, and colonialism*. London: University of Nebraska Press.
Thomas, M., 2014. *Fight or flight. Britain, France, and their roads from empire*. Oxford: Oxford University Press.
Tilly, C., 1984. *Big structures, large processes, huge comparisons*. New York: Russell Sage Foundation.
Triaud, J.-L., 2014. Giving a name to Islam south of the Sahara: an adventure in taxonomy. *Journal of African History*, 55 (1), 3–15. doi:10.1017/S0021853713000820.
Trotignon, 2017. Le Sahel, laboratoire d'un échec contre le djihadisme. *Le Monde*, 22 April.
United Nations, 2015. *DPKO brainstorming brownbag lunch. Countering terrorism and violent extremism: implications for UN peacekeeping*. New York: United Nations.
United Nations, 2018. *Letter dated 8 august 2018 from the panel of experts established pursuant to resolution 2374 (2017) on Mali addressed to the president of the security council*. Official UN document. ONU S/2018/518.
United Nations, 2019. *Situation au Mali. Rapport du Secrétaire général*. Official UN document. ONU S/2019/454.
Vines, A., 2011. Côte d'Ivoire: power gridlock. *The World Today*, 67, 2.

Walker, R.B.J., *et al.*, 2018. Collective discussion: diagnosing the present. *International Political Sociology*, 12 (1), 88–107.

Weber, H., 2007. A political analysis of the formal comparative method: historicizing the globalization and development debate. *Globalizations*, 4 (4), 559–572. doi:10.1080/14747730701695828.

Zobel, C., 2013. Le Mali postcolonial. *In*: Gonin, Kotlok, and Montclos, eds. *La tragédie malienne*. Paris: Vendémiaire, 61–92.

The Middle East is Violence: On the Limits of Comparative Approaches to the Study of Armed Conflict

Jacob Mundy

ABSTRACT

For years, there has been debate as to whether or not the Middle East experiences more armed conflict – and for different reasons – than other regions in the world. Absent is any consideration of the grounds upon which such regional comparisons are possible. Rather than providing a general account of regions, this article instead provides a theory of the Middle East based on the violent practices that have made and reproduced the Middle East as a region, both materially and ideationally. Though critical of comparative approaches to the study of armed conflicts, this argument models a different way to understand them.

'[T]he problem of proper conceptualisation of space is resolved through human practice with respect to it'.

– David Harvey (1973, p. 13; quoted in Paasi 2011, p. 169)

'Geography is therefore the art of war'.

– Edward Said, *Al-Hayat* (10 December 1993)

Introduction

Is the Middle East more violent than any other region in the world?[1] Well before the events of 11 September 2001, this question has provoked debate

This paper's theoretical argument was first presented in an invited talk on 28 September 2017 at Syracuse University's Moynihan Institute of Global Affairs at the Maxwell School of International Studies. The author would like to thank Osamah Khalil for the invitation and for assembling an enthusiastic audience of students and faculty. Written comments from Timur Hammond and Nathalie Koch were particularly helpful in the weeks that followed. A revised version of this paper was presented at the workshop *The Politics of Comparing Armed Conflicts* at the Université du Québec à Montréal on 13 October 2017, an event wonderfully organised by Bruno Charbonneau and Adam Sandor. A Libyan case study was presented at the 2017 annual meetings of the Society for Historians of American Foreign Relations and the Middle East Studies Association. A final version of the paper's theoretical argument, with a specific case study on Western Sahara, was presented at the 2018 World Congress on Middle East Studies in Seville, Spain. My appreciation to Yahia Zoubir and Souad Laghdaf for coordinating the latter. Given the provocative nature of this paper, it goes without saying that all views and any mistakes herein are mine alone
This article has been republished with minor changes. These changes do not impact the academic content of the article.

and polemics from a number of quarters (for background, see Lawrence 1998). Rarely has this question been studied in a systematic and comparative fashion. When it has, most studies have tended to suggest that there is nothing extraordinary about the rates or causes of organised violence in the Middle East in comparison to other regions. The most rigorous effort to compare the levels and aetiologies of armed conflict in the Middle East with other world regions appeared over a decade ago (Sørli *et al.* 2005). It had two relevant conclusions: one, the Middle East is no more conflict prone than any other region in the world and, two, the aetiology of civil wars in the Middle East are no different than those elsewhere. Though now dated and limited in scope to an examination of high intensity internal armed conflicts, this latter finding nonetheless helped to undermine claims that Middle Eastern violence is rooted in regionally specific socio-cultural factors like religion or ethno-linguistic composition (i.e., the predominance of Islamic and Arabic identities), as well as the region's notorious geology (i.e., the predominance of global hydrocarbon reserves in the Persian Gulf, Arabia, Mesopotamia and North Africa). This finding also contributed to a growing consensus among researchers in the first decade of the 2000s that culture and cultural difference (*pace* Huntington 1993) played little role in the causation of civil wars worldwide.

In the years since, serious comparative studies have likewise tended to reject the idea that there is a regionally specific character to armed conflict in the Middle East. A more recent study, for example, suggests that it is a volatile combination of rentier states and authoritarian republics – both being geographical and historical contingencies in the Middle East – that explains much of the region's interstate conflict (Colgan 2013). Similarly, it has been suggested that the continuing prevalence of a particular regime type, monarchies, has in fact had a pacifying effect on the region (Møller 2017). The leading introductory textbook on the political-economy of the region similarly concludes that, in comparison to other regions, there is nothing extraordinary about the number of wars in the Middle East despite the exceedingly militarised nature of many of its regimes (Cammett *et al.* 2015).

Such findings seem difficult to square with recent developments across the wider Middle East in recent years. Of all the armed conflicts to emerge around the world since 2000, over a third have been in the Middle East. Since the 2011 Arab Spring, that number has grown to half. By 2016, the Syrian civil war, then in its fifth year, passed two grim milestones: it had become the deadliest armed conflict of the new century and the second most lethal since the end of the Cold War, just behind the 1994 massacres in Rwanda.[2] Today, four countries in the region – Syria, Afghanistan, Iraq, and Sudan – now rank among the ten most war-ravaged since 1989. From 2012 onward, the Middle East began to eclipse sub-Saharan Africa and Southeast Asia in terms of conflict-related deaths worldwide, a first for the region since the end of the Cold War. The widespread violence across the region that followed the 2011 'Arab Spring' moreover helped to increase the number of ongoing armed conflicts worldwide to unprecedented levels. The fifty-three distinct and active episodes of armed conflict registered in 2016 was a world record for the international order established after WWII (Allansson *et al.* 2017, Pettersson and Eck 2018). As a result of this surge in conflict, half of the recorded terrorism events in the world have likewise occurred in the Middle East since 2011 (START 2018). Thus, despite all the rhetoric and theorization of the post-9/11 counterterrorism initiatives of the United States as a kind of 'everywhere war' (Gregory 2011, Scahill 2013), the geography of this sprawling campaign, whether measured in terms of combatant and civilian fatalities, other humanitarian costs or US military expenditures, is overwhelmingly in and around the Middle East.

Historically, it is also worth noting that a third of all wars between states since 1970 have occurred in the Middle East while nearly forty per cent all internationalised civil wars have occurred there as well during the same period (Sarkees and Wayman 2010; see also Pettersson and Eck 2018). This highly internationalised character of Middle Eastern conflicts is likewise revealed in the extraordinary number of wars of imperial conquest and resistance to occupation in the region. Not only has the Middle East experienced a quarter of all such conflicts over the past two centuries, half of those that have occurred since WWII were in the Middle East. Since 1970, seven of the ten wars of occupation recorded worldwide since 1970 were either in Northern Africa or Southwest Asia, including the invasions of Western Sahara, Afghanistan and Iraq, as well new phases in the Israel-Palestinian conflict (Sarkees and Wayman 2010).

The recent proliferation of full scale wars, low level armed conflicts, seemingly endemic terrorism and foreign military interventions in the wider Middle East, as well as adjacent areas in both Africa (the Maghrib, the Sahel and the Horn) and Asia (the Caucuses and the western Himalayas), raises anew questions concerning the Middle East's putatively unique ability to attract or otherwise generate mass organised violence (e.g., see Lynch 2016, Anderson 2017). These regionally extraordinary levels of violence also call into question the relationship between armed conflict and other putative exceptionalities attributed to the Middle East, whether the prevalence of authoritarianism and the durability of its monarchical regimes (e.g., Bellin 2012); economic issues, such as failed development strategies and squandered oil wealth (Colgan 2014, Ross 2018); or socio-cultural issues, like the treatment of women, repression of minorities, and the growth of religious conservatism (e.g., Olimat 2014, Kymlicka and Pföstl 2014, Cavatorta and

Merone 2017). In many ways, this extraordinary-ordinary debate vis-à-vis the question of Middle Eastern violence has been framed largely in response to Edward Said's (1978) *Orientalism* (see also Lockman 2004, Bonine *et al.* 2011). Extraordinary theories of the Middle East are often decried as neo-orientalist given their implicit or explicit resort to dubious notions of shared culture. Ordinary theories, on the other hand, are effectively distinctions without a difference: why do we need an idea of the Middle East if the Middle East is no different than anywhere else? Hence a whole series of deconstructive questions unfolds, a series that threatens to destabilise the unquestioned yet nonexistent foundations of comparative armed conflict research. If regions are distinction without a difference, then why 'code' for them? If regions matter, then what are they? On what basis can such comparisons even be made within and between them? What are the socio-cultural, political, or geographical properties that naturalise regions and so licence juxtapositions of spatial conglomerations of states, polities, environments and conflicts that are both near and far-flung? What are, in effect, *the conditions of regionality* that unite one region and render it as a coherent unit of analysis, one that can be compared to other regions with similar or different constitutive properties?

In the ongoing debate between extraordinary and ordinary approaches to violence in the Middle East, the very idea of *the region*, in a general sense, goes without any substantive and explicit theorization. The region, as a kind of Bourdieuian (1977) *doxa*, plays a similar function in the wider field of armed conflict research. In the studies and quantitative data cited above, regions are treated as unproblematic and pre-given entities through which analysis can arrive at reasonably sound comparative observations, if not something approaching a statistically derived understanding of likely causal mechanisms. When there is debate about regions, it often takes the form of defining their boundaries rather than rationalising their use. Rarely – outside of the field of geography – is there an effort to come to grips with the ideational and material conditions of *regionality*. The absence of meaningful theoretical engagement with the concept of the region also stands in sharp contrast with the ample debates vis-à-vis statistical methodologies that have come to dominate comparative approaches in the study of armed conflict (Schrodt 2014).

The argument presented here proposes a new way to approach these questions. The question of Middle Eastern conflict – *why does it experience so much violence ?* – cannot be separated from the question of the region itself – *what is the Middle East?* Instead of treating these as separate questions, a more fruitful line of inquiry opens when they are viewed as mutually constitutive. The question of Middle Eastern violence can only be addressed through an examination of the actual processes whereby regions historically emerge, are territorially elaborated and are politically reified as intellectual artefacts, spatial entities, material assemblages and as lived realities. The conclusion arrived at here suggests that the Middle East cannot be understood apart from the intensively violent arrangements, practices and processes that have come to constitute the fundamental conditions of its ideational and material possibility. Efforts to compare historical and contemporary forms of organised violence in Middle East with other regions are thus exposed for their inadequacy. In so far as comparative approaches to armed conflict hypothesise as a possible cause of Middle Eastern violence some innate or unique characteristics of the region, they mistake consequence for condition. In a very real and literal sense, the Middle East is violence.

Contrary to a comparative aetiological approach that problematically attempts to temporally isolate dialectically interrelated phenomena, the kind of *imminent* framework deployed here (Sayer 1987) theorises violence and the Middle East as emerging, existing and evolving together. Though there is some recognition within comparative politics that the independence of cases and causation are analytical fictions (Beissinger 2007; cited in Bellin 2012), an examination of the realisation/regionalisation of Middle East deconstructs, at a more fundamental level, the theoretical assumptions that undergird notions of cases and causation in the first place. Simply put, comparative approaches are incapable of accounting for Middle Eastern violence because they fail to recognise the extent to which that very violence has been ideationally, spatially and materially constitutive of the Middle East. To ask whether or not the Middle East is or is not more violent than any other region is to assume that regions have properties that are ontologically deeper and temporally prior to that violence. What this argument seeks to demonstrate is a more disturbingly dense entanglement of the Middle East and the violence that is said to emanate from it. In pursuing these questions, what is put forward is not so much a theory of Middle Eastern violence as a theory of the Middle East that centres violent practices in the constitution and reproduction of it as a region. Assisting this effort are insights drawn from recent geographical theorizations of the region and territory, poststructuralist accounts of the 'state effect' and its productive indeterminacy, and finally a closer analysis of the ways in which various forms of epistemic and physical violence constitute the very arrangements, practices, and processes that give rise to the 'regional effect' of there being a pre-theoretical and agentic socio-geographical entity called the Middle East. This analysis will examine the ways in which the violence of cartographic practices, imperial-state bureaucracies and academic categorisation have complemented the extraordinary insecuritisation of the Middle East over the course of the previous fifty years through externally driven processes of militarisation, conflict exacerbation and peace deferment. In making the Middle East as violence through violence, the region came to be through its incorporation into larger global processes aimed at restructuring the political and economic dominance of the North Atlantic powers.

Theorising Regions as Political Technologies of Territorial State Space Making

Geographers have long been preoccupied with theorising regions and regionality. As recent works in that field have argued, regions are never given. They are rather constituted, reproduced and contested in numerous sites within and beyond the region, at different spatio-temporal scales and along a number of discursive axes. These dynamics are at once intra-, inter-, extra- and trans-regionally manifold and yet often appear to manifest as regionally specific characteristics. For example, Paasi and Metzger (2017) have argued,

> Work of regionalization and 'region-building' is performed not only by economic, political and cultural/media elites in the production/reproduction of regions and identity narratives, but also in everyday practices and in the work of, for example, regional planners and developers, as well as through such mundane material structures such as transport infrastructures. (Paasi and Metzger 2017, p. 26)

Here our attention is rightly being drawn to local forms of agency in the enactment of regions. At the same time, this observation also exposes one of the more telling features of the Middle East: the Middle East is a region – first in name, then in deed – whose most powerful and devoted stakeholders have never been autochthonous.

The very idea of a Middle East was largely invented out of the ashes of orientalism. It helped to impose an identity upon diverse peoples seeking to forge their own modernity. The idea of 'Latin America', by contrast, is quite instructive: there a regional identity was 'invented' in Southern and Meso America, in part, to resist the very imperialism that had helped to constitute the region and its peoples (Gobat 2013). The sordid career of the Middle East – a concept explicitly rooted in imperial geography, military conquest, political subordination and extractive accumulation, especially since the onset of the hydrocarbon age – has much more to do with the exercise of power over people than with the resistance it generates. The increasing use of the concept of the Middle East from the late 1800s onward coincides with the efforts of North Atlantic powers to dictate the kind of politics the region could have (Lustick 1997, Khalil 2014).

The Middle East as a region must be recognised as a territorial project championed by extraterritorial forces. Territory, according to Elden (2010, p. 811–812), is best conceptualised as a 'political technology', one that 'comprises techniques for measuring land and controlling terrain'. The historical emergence of the region as a technology of rule in modern European geographical thought can be seen in a number of cartographic practices, notably in the latter eighteenth and early nineteenth centuries.[3] These practices included maps that delimited large territorial areas premised on assumptions of relative racial homogeneity. Some of these would later become the names of states; India is a prime example. Other areas or regions were bound together by defunct, decrepit or extant empires. The cartographic territorialisation of regions was complicated by the new wave of European imperial expansion in Asia and Africa in the nineteenth century. The putative ethnological character of emergent understandings of regionality was subordinated to the political imperative to outline, with ever increasing precision, new colonial domains that notoriously disrupted or repurposed the lived geography of those populations being subjected to European domination (Lewis and Wigen 1997, p. 158–162).

The origins and perpetuation of regions through processes of political power discloses the fact that the region is, first and foremost, *state space*. And in so far as the state is the quintessential object of modern politics, the production of state space is interrelated with the processes of capital accumulation and war-making. The ways in which states 'simplify' both representations of space and space itself, so as to enable and enhance the coordination of objects and bodies in the pursuit of power and profit, renders the world in ways that are more intelligible to a variety of modern institutions, above all the military (Foucault 1977, Scott 1998). From the familial household to the colonial slave-plantation to the empire writ large, the violent processes of social reproduction and capital accumulation have gone hand-in-hand with particularly attuned spatial practices (Brown 1995, Mbembé 2003, Owens 2015).

In the early post-colonial world of the mid-twentieth century, the resurgence of regions as a conceptual and practical technology of rule was likewise driven by the requirements of global war-making and capitalist order-remaking on the part of the North Atlantic governments led by the United States. As Lewis and Wigen (1997, p. 163) explain,

> Simply put, the old continental architecture, anchored by the vast category of Asia, proved useless for strategic planning. At the same time, the territorial framework of European colonial empires – the other prominent template for global division – was quickly becoming irrelevant. Ultimately, the result of incremental reforms was a new system of world regions, the basis of postwar 'area studies.'

At the nexus of the US academy and state power, regional geographies were being devised and implemented to facilitate Washington's nascent global hegemony. As with other regions in the embryonic area studies paradigm, Middle East studies in the United States was effectively co-founded by persons whose professional biographies blurred the putative boundaries between defence, diplomacy, intelligence and scholarship (Khalil 2016). For example, the wartime organisation that would later become the Central Intelligence Agency was an important feeder of academic talent in the post-WWII academy, particularly for what would become Middle East studies (Lockman 2016, p. 30). Additionally, the prevalence of anthropologists over geographers within the Ethnographic Board – one of the bodies created in the early 1940s to help the US government understand its new imperial domains – contributed to the highly racialized cartographies of area studies. These cartographies would soon inform the exercise of global power in the emergent Cold War (Lewis and Wigen 1997, p. 162–166).[4] In the years that followed, the reification of regions saw the concept become 'abstract' and 'instrumental': 'A region was now a "data-bound cell" that could be analysed by statistical methods' (Paasi 2011, p. 169). This is an apt description of the problematic way regions are now treated in most comparative studies of armed conflict, as will be seen next.

The Middle East in Comparative Armed Conflict Research

As argued above, the idea of the region developed in both modern political thought and practice as a governmental technology of territorialisation so as to render peoples and places amenable to applications of state power. Direct and indirect forms of cartographic representation of regions became important technologies of truth accompanying these projections, from maps to datasets. None of this occurred without telling complications. Regional boundaries are almost always highly contested, and just as often vaguely demarcated. An understandable response to such contradictions and ambiguities is to reject the region as either a meaningful category of analysis or as representative of an actual lived geography. The problem with this rejection is its failure to come to grips with the extent to which regions, and the Middle East in particular, have been reified into a reality that cannot be so quickly abandoned. In retheorising regions vis-à-vis the practices that constitute them as such, it will be argued that these contradictions and ambiguities, particularly with respect to the Middle East, are in fact an essential component of processes of regionalisation, and so understanding them as such.

One would hardly be surprised to learn that much of the quantitative data used to arrive at conclusions about the patterns of Middle Eastern violence, whether in terms of large-scale international warfare, lower intensity civil conflicts or the conceptually problematic category of terrorism, is riddled with contradictions. The Correlates of War dataset (Sarkees and Wayman 2010), the best historical data on wars in the later modern period, never explicitly delineates the boundaries of the Middle East nor enumerates the states that constitute it. Conflicts in Northern Africa, for example, are mostly included in the Middle East category, such as the brief Algerian civil war of 1962 and the longer one in the 1990s. However, the French conquest of Algeria's central northern region of Kabylia in the 1850s is curiously coded as being in 'Africa', while the 1816 Anglo-Dutch shelling of Algiers is back in 'the Middle East', as are the various other waves of French conquest in Algeria from 1830 onward. The dispute over the former Spanish Sahara at the far western edge of Africa, involving a joint Moroccan and Mauritanian occupation in 1975, is also considered a Middle Eastern war in this data. On the eastern side of the region, all imperial, civil, and interstate wars in Afghanistan are consistently coded as being in 'Asia', whether during the British, Soviet, or NATO periods of occupation. The same goes for all wars in Pakistan since separation from India in 1947. Ottoman wars against Persia are coded as 'the Middle East', though the first civil war between the *Partiya Karkerên Kurdistanê* (PKK, Kurdistan Worker's Party) and the Turkish government (1984–86) is coded as 'Asia', whereas the second is coded as 'the Middle East', as are the two wars involving the PKK and Turkey in Iraq (1991–1992 and 1997). Moreover, the first (1963–1972) and second (1983–1991) wars for South Sudanese independence, as well as the Darfur conflict (2003–2006), are coded as 'the Middle East', though the Ottoman conquest of those same areas in the 1820s is said to be in 'Africa'. Late in the nineteenth century, the joint Anglo-Egyptian suppression of rebellion in Sudan is designated as being in 'the Middle East', whereas the Italians' contribution to the same effort is in 'Africa'. The various phases of civil war in Chad are revealingly assigned to 'Africa', including the most intensive period of Libyan occupation in 1980–1981. The later period of fighting over Chad's northern Aouzou Strip, however, is co-located in two regions as the fighting crossed into Libyan territory – the apparent line between 'Africa' and 'the Middle East' – during the so-called Toyota War of the late 1980s.

Contrary to this somewhat expansive definition of the Middle East, the Uppsala Conflict Data Program (UCDP), which maintains the best data on both high and low intensity conflicts since the end of World War II, utilises a more limited definition of the region. Conflicts in Northern Africa, apart from those involving Egypt, are assigned to the Africa region. Though these instances only appear to account for approximately six per cent of all discrete armed conflicts since 1946 worldwide, the inclusion of cases from North Africa in the Middle East category would actually add over forty per cent more to the Middle East's conflict burden while subtracting eighteen per cent from 'Africa'. On the other side of the region, UCPD opts to treat Afghanistan as part of the Middle East, though the Central Asian republics and, more importantly, Pakistan remain outside of it, even though much of the violence in the latter has been intertwined with the near permanent state of war in Afghanistan since the late 1970s (Pettersson and Eck 2018).

The most comprehensive data on terrorist attacks, the Global Terrorism Database (START 2018), opts for a more expansive understanding of the Middle East, one that incorporates all of North Africa, though terrorist events in Mauritania and Sudan are coded as Sub-Saharan Africa. This data draws the eastern limit of the Middle East at Iran, allotting the Caucus states, the central Asian republics, Afghanistan and Pakistan to different regions. Despite the extent to which terrorism has been increasingly theorised and nominally practiced as a 'transnational' phenomena in recent years, the Global Terrorism Database nonetheless treats incidents that cross regions as occurring in one or the other. That is, there is no option for an event to be in two or more regions. For example, an 3 April 2010, attack on the Algeria border, attributed to Malian 'Tuareg extremists' with an 'unknown' motive, is coded as occurring in 'sub-Saharan Africa'.

This brief survey of major datasets on armed conflict yields two relevant observations. First, there is a widespread insistence on regions, and the Middle East in particular, being an important category of analysis, if not an indispensible one. None of these sources suggest that the Middle East could be simply dissolved into Africa and Asia. Second, there is no agreement as to what countries or limits define the Middle East. Generally speaking, the lack of agreement on definitions of the Middle East tends to elicit two kinds of responses. On the one hand, there are appeals to tradition or pragmatism (e.g., Gritzner 2003, Bonine 2012); on the other hand, there are calls to decolonise understandings of the Middle East through more nuanced representations of the region's peoples and places (e.g., Culcasi 2010). What either approach fails to perceive is the extent to which the Middle East's 'boundary problem' – what constitutes its centre and its limits at any historical moment, including our own – is not a series of contradictions that can be solved through either a lowest-common-denominator delineation, appeals for scientific consensus, or an embrace of historical and cultural complexity. In fact, this boundary problem, and the endless contestation it has licenced, is indicative of the processes that have given rise to the Middle East and reproduced it.

In order to understand how the Middle Eastern boundary problem yields insights into the production of the region in the first place, it is instructive to revisit similar debates from the 1970s and 1980s about the relative autonomy of the state, society, and the economy from each other. In the wake of Althusser's totalising theorization of the state as repressive and ideological apparatuses, questions were raised as to whether or not there was a meaningful distinction to be made between a state, the society it governs, and the economic reproduction of both.[5] In his definitive contribution to this debate, Timothy Mitchell (1991) argues that 'the institutional mechanisms of modern political order are never confined within the limits of what is called the state'. As a consequence, '[T]he boundary of the state (or political system) never marks a real exterior'. Above all, he argues, 'producing and maintaining the distinction between state and society is itself a mechanism that generates resources of power.' He goes on to add, 'The arrangements that produce the apparent separateness of the state create the abstract effect of agency' or a 'state effect' (Mitchell 1991, p. 90–91).

Transposing this argument to the Middle East, if not the idea of regions in general, is simple enough. As an imaginative geography that has never been self-constituted (though is purported to be such), the Middle East represents countless trans-regional 'arrangements'. These arrangements are reproduced in two important ways: one, through the power afforded by insisting on there being a Middle East (and hence an interior and an exterior) such that, two, the Middle East appears to have an existence and capacities independent of, and historically prior to, the very arrangements that realised the Middle East and have reproduce it as such. Notions of the Middle East cannot, long after the fact, be rendered socially or spatially coherent, and so pre-political, by reducing them to either 'the historic core of Islam' (Bonine 2012, p. 98) or what was fashioned by European powers out of the old 'Ottoman-Persian imperial zone' (Harris 2016, p. 10).[6] The only way to unravel the enigma of the Middle East is to consider, in the words of David Harvey (1973, p. 13; quoted in Paasi 2011, p. 169), 'human practice with respect to it'. As will be argued next, the practices that have rendered the Middle East as a zone of extreme violence are themselves uniquely violent, such that the Middle East cannot be theorised in the first place – nor even compared to other regions in the second – without taking into account the very violence that has constituted and reproduced it.

The Middle East as Violent Practices: Mapping, Racialisation and Insecuritisation

So what are the practices that have made the Middle East and, in so doing, imbued it with agentic powers? As will be argued, the evolving transnational arrangements and practices that have effected the appearance of the Middle East have been, at their core, inherently and intensively violent. At the same time, this 'regional effect' and the resultant misprision of consequence for cause are also productive in the very ambiguities and contestations they give rise to by licencing all manner of efforts to know, represent and control the Middle East, efforts that also include resistance from within the region itself. The indeterminacy and contradictions inherent in the ideas and practices of a Middle East are, to borrow from Judith Butler (2006, p. 34), a productive outgrowth of the continual ways in which the arrangements that performatively constitute the Middle East are perceived as forces animated by a seemingly pre-existing, self-regenerating and independently agentic socio-cultural entity, infamous for its oil wealth, authoritarianism, conservatism and protracted, bloody conflicts. Though there are countless ways in which this Middle East has been produced as violence through epistemic and physical violence, three primary mechanisms – mapping, racialisation and insecuritisation – help to elaborate a historical framework to understand the emergence and reproduction of the region. This framework will, in turn, help to elucidate not only the late realisation of the Middle East but also the inherent incongruities in any effort to compare its violence to other regions.

Cartopower

The Middle East emerged at the intersection of two violent practices, capital accumulation and war making, amid the growing crisis of late European imperialism at the turn of the last century. From its inception in British imperialism and its 'popularisation' by US naval college professor Alfred Mahan in a 1902 text on geopolitics and strategy (see Khalil 2016, p. 297, note 4), the idea of the Middle East was a product of imperial-military thought and action, one that was frequently framed from the start as a 'question' or a 'problem', as in a journal article penned by General T.E. Gordon of the British Army in 1900. Unsurprisingly, the first governmental institutions to bear the term 'Middle East' were all in the British Colonial Office (Culcasi 2010, p. 585–586). At the same time, the Middle East of 1903 bears little resemblance to that of 2003. *The Middle Eastern question*, a 1903 book by British journalist and diplomat Valetine Chirol, who helped popularise the term in the early twentieth century, is tellingly subtitled *Or some political problems of Indian defence*. Its original frontispiece was a map spanning eastern Arabia to Tibet and Burma (see Foliard 2017, p. 211). A cursory examination of maps of areas that would come to be called the Middle East, along with maps explicitly claiming to represent the Middle East, reveals the extent to which today's understanding of the region did not coalesce until the mid-twentieth century.

Daniel Foliard's (2017) recent survey of British imperial maps produced between the 1850s through the 1920s, documents the initial emergence of the Middle East as a notion that reflected growing imperial anxieties vis-à-vis post-mutiny India and encroachment from Russia. At the turn of the century, British maps incorporating Arabia, Mesopotamia, the Levant, Turkey, Egypt, and adjacent areas were either labelled 'the Near East' or described the specific sub-regions, nations, states or colonies being documented. As an example of the latter, the map that supposedly made the modern Middle East, the one that bears the 1916 signatures of Mark Sykes and François Georges-Picot, is simply titled *Map of Eastern Turkey in Asia, Syria and Western Persia*. Nonetheless, it was a confluence of events – the First World War, the collapse of the Ottoman Empire and the expansion of North Atlantic dominance into regions that held 'one of the greatest material prizes in world history' – that brought the concept of the Middle East to wider public audiences in the West. At that point, the idea of the Middle East began to be encoded into the both bureaucratic and scholarly discourses concerning areas that had once gone by either many names or the Near East (Foliard 2017, p. 260–261). The years that followed saw the North Atlantic powers, in coalition with the major oil companies of the day, carefully manage the exploitation of the region's oil so as to maintain the industry's profits, often by deliberately retarding the development of the region's industrial, export, labour and political capacities (Mitchell 2011, chapter two; Vitalis 2007).

The Middle East, as an internal relation involving war-making, space-making and global capital accumulation, was further realised in the Second World War. The Middle East had not only become central to the ways in which politicians, militaries and leading industries understood and actualised a world at war with itself (Drysdale and Blake 1985, cited in Sidaway 1994, p. 357), wider publics were incorporated into the Middle East's realisation. A 1941 article in *Time* magazine, for example, titled 'Treasures of the Caucasus', featured a map describing the 'Middle East routes' through which Anglo-American military aid to Russia would have to flow, as well as the natural resources like manganese and oil – 'treasures of the Caucasus' – that 'Hitler covets' (in Rankin 2018, p. 73). Similar concerns animated the realisation of the Middle East during the early Cold War, as the region came to be largely constituted in extrinsic terms vis-à-vis the Soviet threat. Elaborating the Middle East as a unique global space generated by interlocking conflicts and energy extraction coincided with the converging crises that beset the North Atlantic heartlands of capitalism starting in the late 1960s, notably in terms of Keynesian economics (stagflation), conventional military power (Korea, Algeria and then Vietnam), and political legitimacy (new social movements). It is at this point in recent history when we can begin to speak of the Middle East in terms that would be recognisable today. At the same time, this new Middle East had to be re-rationalised in such a way as to justify its growing incorporation into various schemes aiming to shore up North Atlantic political and economic hegemony. This justification would largely take the form of racialising the Middle East so as to naturalise cartographies that had previously been explicitly geopolitical.

Racialising the Region

The gradual racialisation of the Middle East – replacing its political-economic logic with a socio-cultural one – coincided with the region's reorientation westward towards zones of hydrocarbon energy extraction in Mesopotamia, Arabia and North Africa. In the postwar context of US intellectual and political supremacy, new ideas of the Middle East had to contend with the legacies of orientalism and other forms of European imperial science. These were largely philological ways of knowing that heavily informed the nascent field of area studies (for background, see Lockman 2004, Kuper 2005, Mamdani 2012). The most obvious legacy was the very idea of a 'middle' in 'the East'. As noted above, the Middle East, in its original formulation in the first years of the twentieth century, was British India, not southwest Asia (Sidaway 1994). Areas adjacent to British India – Afghanistan, Qajar Persia, and even western China – were sometimes included in this early definition. The concept of Near East, by contrast, had signified what was left of the Ottoman Empire in the late 1800s, as well as nearby areas in Southeast Europe, the Levant, Mesopotamia, Arabia, Egypt, and some – or all – of the Maghrib. It would take two world wars and the anxieties of the emerging post-imperial international system to produce a new understanding of the Middle East. Though the US State Department continued to use the term Near East, John Foster Dulles, the agency's Secretary in 1957, declared the two regions 'identical' (quoted in Culcasi 2010, p. 586; citing Davison 1959). Rather than being a product of the First World War, as is commonly suggested, the contemporary Middle East began to be realised during the Cold War.

This reorientation of the Orient, whereby the *Near* became or merged with the *Middle* (see Khalil 2014, Foliard 2017), was not the only source of confusion and contestation during the early Cold War. There was a related challenge at the level of academic institutions. As postwar social sciences were increasingly tasked with going beyond Europe to explore, explain and so help manage the totality of the human world, articulating a workable concept of the Middle East had to confront three additional challenges. It first had to contend with the intellectually and institutionally entrenched philological traditions of Orientalism as enshrined in departments of Near Eastern Languages and Civilisations (Lockman 2016).

Secondly, the new idea of the Middle East had to contend with an undecided division of labour between a nascent Middle East studies and the continental regions of the world from which the Middle East would be carved out: Asia and Africa. As a concept, the continent's fictive geological naturalness in modern European geographical thought (see Lewis and Wigen 1997) gave it a clear advantage over the Middle East, a notion that was not only *a priori* Eurocentric but also *a priori* political, and hence unnatural. These theoretical and disciplinary tensions were indicative of longstanding tensions within the 'modern geopolitical imagination' between natural and constructed notions of space (Agnew 2004). From the perspective of a scientific worldview that valued the pre-given nonhuman world over the ideological and constructed one of the socio-political, an area studies paradigm rooted in the continent – Africa, Asia, Latin America – held certain advantages.[7] A telling fact is the delayed establishment of the Middle East Studies Association (MESA), the largest professional body of scholars working in this new field. It was founded almost a decade after the Sputnik crisis, the event that led to the US National Defence Education Act (NDEA) of 1958 and the dramatic expansion of government funding for area studies in the United States. By contrast, the North American scholarly associations for Asian and African studies had been founded before the NDEA, in 1956 and 1957 respectively (Mitchell 2002b). Up until that point, Middle East studies had been simultaneously facilitated by a coalition of private and public interests within the emergent US national security state, the Middle East Institute being a prime example (Khalil 2016). Though area studies is often criticised for being an instrument of the US national security state (see Sidaway 2013, Koch 2016), the founding of MESA is indicative of an effort of scholars to privatise that which was more overtly under government control.

A third challenge to the idea of the Middle East took the form of efforts to create alternative nomenclatures. The most significant was efforts to promote Southwest Asian studies. However, these never came close to contending with the popularity of 'the Middle East', if, for starters, because of the 'problem' of Egypt and the 'Arab' Maghrib. The use of the term Southwest Asia by diplomats and mapmakers on government documents before and after the interwar period (see, e.g., Foliard 2017, p. 224, Bonine 2012, p. 64–65, Culcasi 2010, p. 586) is suggestive of the extent to which a more socio-politically neutral alternative was readily available yet never found purchase.

More importantly, the curious insistence that there was a Middle East and that it had to be called such spoke to the persistence of racial logics at work in both postwar international thought and the rise of area studies (Vitalis 2015). Regions and their associated identities had to conform to a rigid, arbitrarily imposed socio-geographical schema, one hostile to hybridity and interpenetration, yet whose boundaries never elicited scholarly or political consensus (Paasi 2003). What is remarkable about the idea of the Middle East is not only its westward drift in the first half of the twentieth century but its corresponding racialisation. Whereas early formulations of the Middle East included countries like Sri Lanka, Ethiopia and Greece well into the 1950s (Khalil 2014), most maps of the region today attempt to circumscribe an area limited to areas of 'brown' Islamic majorities. Where ambiguities of delineation begin to emerge often appear in zones where this purported religious and racial preponderance begins to decline in Africa, central Asia, and southeastern Europe.

Maps of the Middle East reflect the region's racialisation. In a chapter on the seemingly irreconcilable maps of the Middle East found in post-WWII geographical texts, Michael Bonine (2012) examines the implicit and explicit logics at play in any given definition of the Middle East so as to arrive at a common basis – the Middle East as a 'cultural region' – that can then rationalise cartographic representations of it.[8] However, were we to analyse these same maps diachronically, a different trend comes into focus. The maps of the Middle East produced either right before the Second World War or in the years immediately afterwards hew closely to the old Near East or Southwest Asia understanding. By the 1960s, these maps begin to expand the scope of the Middle East, mainly into Northern Africa. Even maps that attempt to define the region according to geological features or climatological patterns just as much reflect the expanding concept of Middle East in relation to the geopolitics of the day. One of the most obvious aspects of this sprawl is Libya, which begins to be incorporated into maps of the Middle East as the country becomes a 'problem' for North Atlantic powers in the 1970s. By the 1980s, most or all of the Maghrib states begin to appear in these maps of the Middle East, some of them now bearing the title 'the Middle East and North Africa'. Nearly all of the maps produced after 11 September 2001, in Bonine's study not only include the Maghrib, but several suggest the inclusion of Afghanistan, Pakistan, and the central Asian republics into concepts of a 'wider' Middle East. At the far west of this sprawling Middle East, Mauritania and the contested territory of Western Sahara continue to present an epistemic challenge, oscillating between incorporation and 'Africa'. Tellingly, South Sudan's independence has made it easier to render Sudan as Middle Eastern. At the same time, increasing North Atlantic counterterrorism activities since 2002 in Africa's Sahel belt – Mauritania, Mali, Burkina Faso, Niger and Chad – have further pushed the scope of the Middle East's disputed liminal zones further south. This expansive imaginative geography is nearly coterminous with the primary theatre of operations in the sprawling Global War on Terrorism (Güney and Gökcan 2010).[9] Thus the realisation of the Middle East through its increasing racialisation plays an important role in simultaneously masking the actual violent practices that have made the Middle East while also making that violence possible by 'othering' the Middle East as a race-based space of culturally inherent authoritarianism and conflict.

Insecuritisation as Accumulation

The contemporary Middle East began to be realised and consolidated, both ideationally and materially, as race-based space in final third of the twentieth century. This occurred amid the growing crises of politics and economics afflicting the North Atlantic powers towards the end of the 1960s as Western capitalism and military might increasing came into question. A partial and contingent solution to some of these crises was found in the Middle East by making it what it has become over the course of the previous fifty years: a unique zone of violence. Driven by militarisation, conflict exacerbation and peace deferment, these policies of insecuritisation helped to generate new modes and spaces of capitalist accumulation for beleaguered North Atlantic arms and oil industries. Making and maintaining the Middle East in terms of permanent war and irresolvable conflicts was accomplished through processes of epistemically violent realisation. These processes not only took the form of intellectual practices in both the public and private domains (e.g., government bureaucracies and Middle East studies) but, in so doing, they also helped to structure and legitimate the region's realisation through the physical violence of hyper militarisation, conflict exacerbation and peace deferment. The extraordinary and intensive insecuritisation of the Middle East in the 1970s helped to actuate this co-constitutive relationship, a relationship between mass organised violence, accumulation through insecurity and new racialised understandings of the region itself.

Prominent critics of US policy towards the Middle East have often suggested that the region held immediate and profound geopolitical importance to North Atlantic interests from the interwar period to the early Cold War (e.g., Klare 2004, Bacevich 2016). As evidence, such critics cite US government support for Standard Oil of California to explore oil in Arabia (at a time when the United States was still a net exporter), the 1945 meeting between President Franklin D. Roosevelt and King Ibn Saud on the Great Bitter Lake in the Suez Canal or government documents like the 'Acheson Memo' of October 1945, which infamously described Arabian oil reserves as 'a stupendous source of strategic power, and one of the greatest material prizes in world history'. Moreover, the 'Truman Doctrine', effectively promulgated in 1948, is also presented as an indication of a strong and early US commitment to pre-eminence in the Middle East. The historical record, however, is far more equivocal as to the veracity of this geopolitical thesis. For example, key cases like the early Saudi state or the founding of Israel raise questions as to whether or not the real advocates for US entanglement in the region came from the region itself (see, respectively, Vitalis 2007, Gendzier 2015). From a political-economy perspective, the Middle East effectively became central to the interests of the North Atlantic world in the late 1960s and early 1970s, and for reasons that were quite different than the geopolitical context of the interwar period or the early 1950s. The crisis of North Atlantic hegemony in the late 1960s coincided with the very moment, as argued above, that the contemporary Middle East came to be. What the Bitter Lake meeting, the Truman Doctrine and the Acheson Memo provide is a trite political origins story that obfuscates the actual political-economy of the growing North Atlantic-Middle East entanglement of the last fifty years.

In two important ways, insecuritising the Middle East in the 1970s – and, in so doing, producing the region as it is understood today – helped to address twin crises afflicting North Atlantic capitalism and military power. First of all, insecuritising the Middle East helped to set in motion an interactive chain of often-violent antagonisms stretching from the western edge of the Sahara desert to the western Himalayas. The most common logic at the root of all of these antagonisms was not some violent cultural habitus but rather essentially internationalised political struggles to define and control the Middle East.[10] Given the growing worldwide opposition to

North Atlantic military interventions in the postcolonial world in the 1950s and 1960s, producing the Middle East as an archipelago of 'internal' or 'regional' conflicts afforded a degree of indirect control over a zone that was increasingly obtaining the financial and ideological resources to forge its own political modernity during the same period. Secondly, insecuritising the region helped to address crises of profitability in both the North Atlantic's armament industry and hydrocarbon-based energy sector. In both economic and political terms, the Middle East served as a productive 'spatial fix' (Harvey 2001) so as to 'buy time' (Streeck 2014) for key sectors within North Atlantic capitalism (arms and oil). This occurred amid the ongoing project to reconstitute North Atlantic hegemony after successive disasters in a number of postcolonial military theatres (e.g., Korea, Algeria, Kenya and Vietnam). And as much as a solution to these crises of power and profit were found in the Middle East, they were effective in so far as they were able to produce the Middle East as a space of violence through violence.

The foundations for the intensive and externally driven insecuritisation of the Middle East were laid in the 1970s through the region's hyper militarisation. Recent scholarship on the problem of militarism has rightly framed it as a relationship between the 'legitimation', 'normalisation' and 'preparation' for war (Stavrianakis and Stern 2018). These facets of Middle East militarisation are already well recognised, both in terms of the high levels of security sector spending by the states in the region and the amount of military aid they have received from governments outside of it (Cammett *et al.* 2015). Historically, many of the regimes in the region, dominated both ideationally and institutionally by their security sectors, were often the product of colonial-military administrations (Vitalis and Heydemann 1999), or a confluence of private and strategic interests in cases like Saudi Arabia (Vitalis 2007), Iran (Alvandi 2014) and Libya (Anderson 1986), where the creation of Western-dependent security and extractive apparatuses were prioritised above all other spheres of governance. Even in cases where European imperialism was shed through violent resistance, such as Algeria (Roberts 2007) or Egypt (Abul-Magd 2017), the postcolonial state that emerged was nonetheless heavily influenced by military elements drawn from the nationalist insurgency, indigenous forces that had defected from the colonial army or a mix of the two.

On their own, the institutional dominance of security sectors within Middle Eastern regimes and the developmental logic of 'military Keynesianism' are unable to explain the extraordinary levels of military spending and imports across the region. As indicated in the stunning number of arms imports in the 1970s, these militaristic tendencies were exacerbated by a confluence of forces exogenous to the region. In 1972, the Middle East had only accounted for

19.2% of world arms imports with East Asia (37.7%) and Europe (28.2%) then dominant because of the Vietnam war and Cold War arms build-ups (US Arms Control and Disarmament Agency 1984, p. 7, Table C). Whereas the Middle East collectively spent $1.26 billion on arms in 1963, a decade later its expenditures were already over $10 billion in unadjusted dollars. This reflected an annual rate of growth of 14.7%, far ahead of East Asia (7.5%), South Asia (2.9%), Africa (6.5%), and Latin America (3.9%) (US Arms Control and Disarmament Agency 1975, p. 14–19). The intensive militarisation of the Middle East in the 1970s took place in a context where arms sales worldwide grew from $300 billion in 1972 to roughly $820 billion ten years later. The Middle East accounted for much of this growth: in 1979, the Middle East received 34% of global arms deliveries. Though this was slightly down from 1978, it still represented a figure 3.5 times larger than it had been in 1970 (US Arms Control and Disarmament Agency 1982, p. 5, 9 and figure 5). These trends continued well into the initial years of Thatcher premiership and Reagan administration. In 1982, almost 42% of all global arms transfers went to the Middle East, which was roughly half of the developing world's arms imports (US Arms Control and Disarmament Agency 1984, p. 1). In 2017 dollars, Middle East arms imports went from below $7.5 billion in 1971 to over $18 billion in 1975, and then to over $30 billion in 1977 (US Arms Control and Disarmament Agency 1982, p. 8 and figure 13). By the end of the decade, the countries whose military expenditures registered as 10% or more of GDP were almost all in the Middle East apart from the Soviet Union – both Yemeni states, Egypt, Jordan, Oman, Syria, Qatar, Israel, and Saudi Arabia. The ten countries with the highest ratio of military expenditures to central government expenditures in 1982 were all in the Middle East as well (US Arms Control and Disarmament Agency 1984, p. 5, Table B).[11]

These trends have continued well beyond the Cold War. Though average military spending across the Middle East peaked at close to twenty per cent of GDP in the early 1980s, most of the region's states have nonetheless continued to outpaced all other 'middle income' countries world-wide in terms of military spending and arms imports since the collapse of the Soviet Union. These figures include low population oil-exporters and high population countries with limited natural resources. Unsurprisingly, these spending trends have been dominated, especially since 2001, by the states of Saudi Arabia, Israel, Turkey, the United Arab Emirates and Iran. Some of these states saw acute bursts of exorbitant military outlays during times of war while others more consistently propped up elaborate internal security forces. Meanwhile, foreign military sales and aid to the region, led by the North Atlantic powers and Russia since the 1970s, has never abated, outpacing all other regions combined. Domestic funds and rents that otherwise could have gone to sectors like education, health, and civil infrastructure were massively diverted to defence, which in turn empowered, through enhanced surveillance and repressive capacities, the authoritarian tendencies of regimes dominated by the logic and institutional structures of disproportionately large security sectors (Cammett *et al.* 2015, p. 357–366; see also Bellin 2004).

Insecuritisation through militarisation in the 1970s helped to make the Middle East through violence by actuating an interlocking chain of conflicts from Mauritania to Pakistan.[12] How these productively destabilising dynamics have emerged in the 'core' – the Arab-Israeli and Israel-Palestinian conflicts, civil wars and foreign occupation in Lebanon and Yemen, the Iran-Iraq war, the Kurdish question, the Gulf Wars, the pariahisation of Iran and finally the Syrian civil war – are all well understood. Equally part of this emergent system in the 1970s were new conflicts on the 'margins' of the Middle East that likewise helped to constellate and maintain an archipelago of regional insecurity through the induction of sub-regional antagonisms, the exacerbation of civil and international wars, the selective pariahisation of particular regimes and the prolongation of intractable 'peace processes'. On the western wing of the region emerged the Saudi-financed Moroccan occupation of Western Sahara, which was opposed by Algerian backed nationalists (still ongoing), and the Soviet-armed Libyan war in Western-backed Chad. Libya's *de jure* pariah status and Algeria's *de facto* one later resulted in widespread intellectual and political paralysis vis-à-vis the civil wars that erupted in the latter in the 1990s and the former following the 2011 NATO-Arab League intervention. Preemptive counter-terrorism initiatives in North Africa's Sahel belt from 2002 onward only served to further destabilise the poorest region in the world, leading to a proliferation of armed 'jihadi' groups from Niger to Mauritania, the collapse of the Malian state in 2012 and the rise of 'Islamic states' in Timbuktu and the Lake Chad region.

At the other end of the Middle East is a conflict that needs even less of an introduction: the four-decade evisceration of Afghanistan based on the hyper militarisation of a Cold War ally (Pakistan), Saudi financing and North Atlantic intervention. By 2014, the United States had spent more money on security assistance and nation-building in Afghanistan since 2001 than was spent on the Marshall Plan for the reconstruction of Europe after WWII. The expert consensus nonetheless suggests that Afghanistan remains a failed state mired in an intractable conflict nearly two decades after 9/11 (Messner 2018). The conflicts over Palestine, Western Sahara and Afghanistan (recent negotiations notwithstanding) furthermore highlight the extent to which preventing peace has likewise been an important mechanism of insecuritisation in the Middle East. Data on over two hundred peace agreements reached between 1975 and 2011 around the world shows that less than ten per cent pertain to conflicts in the Middle East, the majority of those being a series of agreements between Iraq and Iran and the various phases of the Oslo peace process (Högbladh 2012).

Thus the articulation and maintenance of interlocking yet productive instabilities from the western Sahel to the western Himalayas from the 1970s onward has combined hyper-militarisation, conflict exacerbation, deferment of peace and, at times, direct intervention. By hijacking the logic of the Cold War and all-the-while maintaining it long after the fall of the Berlin Wall, these manifold processes of regional insecuritisation were abetted by geographically patterning the region into an oscillating arc of North Atlantic enemies – Algeria, Libya, Egypt (until 1973), Syria, and Iran (after 1979) – and allies – Morocco, Tunisia, Israel, Saudi Arabia, Iraq (until 1990) and, until recently, Pakistan and Turkey. Aiding and abetting these processes was the coincidental invention and development of a new framework of understanding and so managing organised violence in the Middle East, particularly by nonstate actors, as 'terrorism' (Zulaika and Douglass 1996, Stampnitzky 2013). Discourses and practices of (counter)terrorism likewise facilitated the reproduction of the Middle East through violence by – first and foremost – rendering as much indigenous violence as *a priori* illegitimate and imminently menacing. The result is that interventions against terrorism obtain the status of being simultaneously legitimate yet hazardous, a status with clear political urgency and strategic flexibility.

In structuring this chain of antagonistic relations, armed conflicts and hyper militarised regimes, the insecuritisation of the Middle East from the 1970s onward helped to reformulate North Atlantic political supremacy through more indirect forms of control over an increasingly divided postcolonial world. These processes also functioned as new mechanisms of capital accumulation so as to address growing crises within Western arms and oil industries. These crises were respectively rooted in, one, potential US arms spending reductions as the Vietnam war came to a close and, two, the wave of infrastructural nationalisations and profit-sharing renegotiations that had undermined the cartel power of the major North Atlantic oil companies (Mitchell 2002a). In order to solve these crises, new markets for arms would need to be opened and a new regime of managed oil scarcity would have to be devised.

On both fronts, Middle Eastern insecurity proved exceptionally capable of satisfying these accumulatory needs. In the wake of the 1973 Arab-Israeli war and the Arab oil 'embargo', it soon became clear that widespread insecurity in the Middle East – where seventy per cent of the world's proven oil reserves reside – could manufacture an aura of oil scarcity that had previously been effected through the monopolistic powers of the major international oil companies. Skyrocketing oil prices in the 1970s, driven by unreliable pariahs, militarisation, wars and a proliferation of 'intractable' conflicts across the Middle East, from West Africa to Central Asia, helped to resecure the dominance of the international oil companies. The share of profits going to the Middle Eastern states was then 'recuperated' by the North Atlantic powers via extravagant arms sales to an increasingly insecuritised region. Simply put, the relative or 'differential' profitability of the major North Atlantic arms and oil companies (i.e., in comparison to other dominant sectors within global capitalism) became predicated on the initiation and perpetuation of major wars and intractable conflicts in the Middle East in the 1970s (Rowley *et al.* 1989, Nitzan and Bichler 2002, chapter five). The elegance of this system, which induced and perpetually exacerbated vicious internal and external security dilemmas for almost all states in the region, rested in its ability to incorporate countries with no direct access to oil financing (e.g., Israel, Pakistan and Morocco) and petro-pariahs armed with Soviet/Russian weapons (e.g., Algeria, Libya, Syria, and post-1979 Iran). Nonetheless, contradictions emerged within this system, mainly in the form of new incentives to expand the geographical and geological frontiers of oil extraction such that extended periods of 'overproduction' proved difficult to reverse. Though the problem of the oil 'glut' of the mid 1980s and the 1990s was resolved with the onset of the post-9/11 wars in Afghanistan and Iraq, conflicts that temporarily restored the relative profitability of the dominant North Atlantic oil and arms corporations (Bichler and Nitzan 2015), the contemporary oil glut from 2014 onward – induced by the dramatic expansion of oil extraction in the United States – has so far been difficult to clear even with major production zones and circulation corridors under constant threat (e.g., Libya and the Straight of Hormuz) and two major producers, Iran and Venezuela, under increasingly tight unilateral US sanctions regimes put in place by the Trump administration. Nonetheless, it is now plainly clear that, over the course of the previous fifty years, Middle Eastern insecurity has become as much an important condition of possibility for the realisation of profitability within key energy and arms industries as this insecurity has become a mechanism through which North Atlantic political supremacy has been maintained.

Conclusion

To say that the Middle East is violence is to recognise the extent to which the Middle East is unthinkable apart from the forms of ideational and physical violence that have come to constitute it, both as territory and within the contemporary political imagination, as a race-based 'space of exception' premised on violence. It is also to suggest that efforts to compare war, conflict, terrorism and other forms of organised violence in the Middle East with similar forms of violence in other regions is to commit a fundamental error. This error arises from the fact that such comparisons mistake consequence for cause. Comparative approaches to the study of armed conflict search for the origins or correlates of organised violence in an entity – the Middle East – that is in fact constituted by that very violence, among myriad other practices that have created this effect. Here it has been argued that an actual understanding of the relationship between organised violence and the Middle East must first attend to the arrangements, practices and processes that have given rise to this regional effect; that is, the actions that create and reproduce the appearance of the agentic socio-geographical phenomena called regions. As with the concept of the state, which must likewise be approached through an analysis of the practices that constitute and reproduce it, the contestation and ambiguity that surround efforts to define and control regions are indicative of such performative origins.

The Middle East, a notion born amid the anxieties of late British imperialism in India, only came to fruition during the latter half of the twentieth century. This is an important observation because it temporally locates the ideational and material consolidation of the contemporary Middle East at a moment of acute crisis in the North Atlantic world. It was during this period when the Middle East's strange imperial origins were obfuscated by its reinscription as race-based space. That this new 'cultural' understanding of the Middle East yielded inconsistent delineations and endless contestation was reflective of the widespread misinterpretation of the Middle East as a socio-geographical object existing prior to the transnational arrangements, practices and processes that have given rise to this regional effect and perpetuated it. Chief among these practices, as argued above, was a convergence of extra-regional political and economic crises whose solutions were to be found through the pervasive insecuritisation of the Middle East. The key insecuritising mechanisms identified above were hyper militarisation, conflict exacerbation and peace deferment, arrangements that helped to pattern the Middle East into an archipelago of interconnected antagonisms. In this way, deepening levels of Middle Eastern insecurity assumed an important global function in the reproduction of North Atlantic political and economic supremacy from the 1970s onward. That organised violence in the Middle East has achieved record levels in recent years is as much a testament to the ability of these powerful arrangements to reproduce and amplify themselves as it is to the specifically violent functions the Middle East serves in the ongoing global contest to forge the future of modernity.

Notes

1. Here violence and organised violence are being used as shorthand, albeit highly problematic, for what Charles Tilly (2003) might have simply described as coordinated acts that cause damage to peoples and their environments. The use of this broad categorisation, which seeks to incorporate civil wars, lower level armed conflicts, terrorism and civilian-directed atrocities, is also meant to align with Uppsala Conflict Data Program's recent shift in attention from 'armed conflicts' to 'organised violence' more broadly.
2. This claim, of course, is not without controversy. According to the Uppsala Conflict Data Program (Allansson *et al.* 2017) the Syrian civil war had claimed

over 280,000 *direct* fatalities by the end of 2016; that is, fatalities in contexts that are effectively intentional acts of homicide committed by state and non-state actors against combatants and civilians (e.g., combat and massacres). It is commonly reported that the violence in the various conflicts in the Democratic Republic of the Congo (DRC) since 1998, especially the Second Congo War or Africa's 'World War' (1998–2003), have claimed well over two million lives. For example, the International Rescue Committee, which conducted a number of excess mortality surveys in the country, claimed in 2008 that as many as 4.53 million had died in the DRC since 1998. These were deaths caused *indirectly* by the conflict. Challenging this figure, the 2009 Human Security Report questioned the International Rescue Committee's methods and findings, proposing instead a 'best estimate' of 863,000 excess (indirect) deaths. The Uppsala Conflict Data Program conservatively lists a total of 102,192 *direct* fatalities in armed conflict in the DRC from the end of the Cold War to 2016. Similar methodological controversies have surrounded efforts to approximate the number of direct and excess deaths in Iraq during the sanctions regime of the 1990s and following the 2003 US-led invasion (e.g., a 2006 study published in the *Lancet* controversially reported over 600,000 direct deaths in armed violence and over 650,000 indirect deaths between March 2003 and mid-2006). On the political and methodological issues of using direct fatalities as a measure of conflict duration, intensity, and typology, see Ghobarah *et al.* (2003) and Mundy (2011).

3. For background on the 'cartographic revolution' in early modern European statecraft, see Branch (2013).
4. Not to be outdone, the Soviet Union would likewise embark upon a massive project to map the world in excruciating detail, one that would also include the intensive mapping of peoples based on religion, language and other markers (see Davies and Kent 2017).
5. Often called the Miliband-Poulantzas debate. For an introduction, see Miliband (1977), Poulantzas (1978), and Abrams (1988).
6. Harris (2016, p. 21) later describes the 'MENA' as having become a 'world capitalist region' after WWII. This categorisation would be more consistent with this paper's argument.
7. The simple solution to this problematic bifurcation – natural/unnatural, given/constructed – would have been to call into question the socio-politically manufactured characteristics of all spatial constructions, which would suggest that the regional concept of the Middle East was no more fictitious or unnatural than the continental concept of Eurasia or Australia. This theoretical revolution, however, would only come about in the 1970s in the fields of geography and urban studies (see Harvey 1973, cited in Paasi 2011, p. 169, LeFebvre *et al.* 2003, 2009), as well as, and perhaps even more infamously, Middle East studies (Said 1978), more of which will be said below. On the relationship between 'modernity' and 'nature', see Latour (1993).
8. Culcasi (2010) performs a similar analysis on a sample of atlases, textbooks, monographs, and other high-level scholarly sources. She produces a map of where there is consensus on the Middle East 'core' and the 'periphery' so as to arrive at a possible natural or social reason to constellate all of these lands into a 'world cultural region', though she ultimately finds that a consistent and uncontested criteria ('uniform characteristics') cannot be arrived at. Her conclusion is that the Middle East's 'Western geopolitical origins' has to be

critically engaged so as to address the relationship between 'oversimplified notions' and the 'increased involvement' of foreign powers, namely the United States, in the region (Culcasi 2010, p. 590–594).

9. The absurdity of this sprawl is perhaps best witnessed in the almost neo-Mackinderian vision of post-9/11 geopolitics popularised by Thomas Barnett (2004).
10. Ian Lustick (1997) traces these patterns – North Atlantic efforts to subvert the rise of regional 'hegemons' – from the late Ottoman period through the first Gulf War. Mitchell (2002a, 2011) discusses the ways in which various alliances between North Atlantic powers and Middle Eastern regimes, despite their seemingly contradictory ideological character, have been necessary for both the advancement of global capitalism (particularly after the crises of the 1970s) and the maintenance of US hegemony, albeit both at expense of regional solidarities.
11. In these figures provided by the US government, the 'Middle East' excludes the states of the Maghrib minus Egypt, which are instead categorised in the Africa region. In North Africa, even the resource-poor agricultural state of Morocco was spending somewhere between 5% and 10% of its GDP on arms in the late 1970s, largely as a result of the belligerent occupation of Western Sahara, whereas oil-rich Algeria and Libya were spending between 2% and 5% (US Arms Control and Disarmament Agency 1982, p. 6 and figure 9).
12. Though this might seem like an untenably expansive definition of the Middle East, it is actually on par with South America in terms of land size or sub-Saharan Africa in terms of population.

Disclosure statement

No potential conflict of interest was reported by the author.

ORCID

Jacob Mundy http://orcid.org/0000-0003-4166-7736

References

Abrams, P., 1988. Notes on the difficulty of studying the state (1977). *Journal of Historical Sociology*, 1 (1), 58–89. doi:10.1111/johs.1988.1.issue-1.

Abul-Magd, Z., 2017. *Militarizing the nation: the army, business, and revolution in Egypt*. New York, NY: Columbia University Press.
Agnew, J., 2004. *Geopolitics: re-visioning world politics*. New York, NY: Routledge.
Allansson, M., Melander, E., and Themnér, L., 2017. Organized violence, 1989–2016. *Journal of Peace Research*, 54 (4), 574–587. doi:10.1177/0022343317718773.
Alvandi, R., 2014. *Nixon, Kissinger, and the Shah: the United States and Iran in the Cold War*. New York, NY: Oxford University Press.
Anderson, L., 1986. *The state and social transformation in Tunisia and Libya, 1830–1980*. Princeton, NJ: Princeton University Press.
Anderson, S., 2017. *Fractured lands: how the Arab world came apart*. New York, NY: Anchor Books.
Bacevich, A.J., 2016. *America's war for the greater Middle East: a military history*. New York, NY: Random House.
Barnett, T.P.M., 2004. *The Pentagon's new map: war and peace in the twenty-first century*. New York, NY: Putnam.
Beissinger, M., 2007. Structure and example in modular political phenomena: the diffusion of bulldozer/rose/orange/tulip revolutions. *Perspectives on Politics*, 5 (2), 259–276. doi:10.1017/S1537592707070776.
Bellin, E., 2004. The robustness of authoritarianism in the Middle East: exceptionalism in comparative perspective. *Comparative Politics*, 36 (2), 139–157. doi:10.2307/4150140.
Bellin, E., 2012. Reconsidering the robustness of authoritarianism in the Middle East: lessons from the Arab Spring. *Comparative Politics*, 44 (2), 127–149. doi:10.5129/001041512798838021.
Bichler, S. and Nitzan, J., 2015. Still about oil. *Real-world Economics Review*, 1 (70), 49–79.
Bonine, M.E., 2012. Of maps and regions: where is the geographer's Middle East? *In*: M.E. Bonine, A. Amanat, and M.E. Gasper, eds. *Is there a Middle East?: the evolution of a geopolitical concept*. Stanford, CA: Stanford University Press, 56–99.
Bonine, M.E., Amanat, A., and Gasper, M.E., eds., 2011. *Is there a Middle East? The evolution of a geopolitical concept*. Stanford, CA: Stanford University Press.
Bourdieu, P., 1977. *Outline of a theory of practice*. Translated by R. Nice. New York, NY: Cambridge University Press.
Branch, J., 2013. *The Cartographic state: maps, territory, and the origins of sovereignty*. New York, NY: Cambridge University Press.
Brown, W., 1995. *States of injury: power and freedom in late modernity*. Princeton, NJ: Princeton University Press.
Butler, J., 2006. *Gender trouble: feminism and the subversion of identity*. New York, NY: Routledge.
Cammett, M., *et al.*, 2015. *A political economy of the Middle East*. Boulder, CO: Westview Press.
Cavatorta, F. and Merone, F., eds., 2017. *Salafism after the Arab awakening: contending with people's power*. New York, NY: Oxford University Press.
Colgan, J.D., 2013. *Petro-aggression: when oil causes war*. New York, NY: Cambridge University Press.
Colgan, J.D., 2014. Oil, domestic conflict, and opportunities for democratization. *Journal of Peace Research*, 52 (1), 3–16. doi:10.1177/0022343314545526.
Culcasi, K., 2010. Constructing and naturalizing the Middle East. *Geographical Review*, 100 (4), 583–597. doi:10.1111/gere.2010.100.issue-4.

Davies, J. and Kent, A.J., 2017. *The Red Atlas: how the Soviet Union secretly mapped the world*. Chicago, IL: University of Chicago Press.

Davison, R.H., 1959. Where is the Middle East? *Foreign Affairs*, 38 (4), 665–675. doi:10.2307/20029452.

Drysdale, A. and Blake, G.H., 1985. *The Middle East and North Africa: a political geography*. New York, NY: Oxford University Press.

Elden, S., 2010. Land, terrain, territory. *Progress in Human Geography*, 34 (6), 799–817. doi:10.1177/0309132510362603.

Foliard, D., 2017. *Dislocating the Orient: british maps and the making of the Middle East, 1854–1921*. Chicago, IL: University of Chicago Press.

Foucault, M., 1977. *Discipline and punish: the birth of the prison*. New York, NY: Vintage Books.

Gendzier, I.L., 2015. *Dying to forget: oil, power, Palestine, and the foundations of U.S. policy in the Middle East*. New York, NY: Columbia University Press.

Ghobarah, H., Huth, P., and Russett, B., 2003. Civil wars kill and maim people, long after the fighting stops. *American Political Science Review*, 97 (2), 189–202. doi:10.1017/S0003055403000613.

Gobat, M., 2013. The invention of Latin America: A transnational history of anti-imperialism, democracy, and race. *The American Historical Review*, 118 (5), 1345–1375. doi:10.1093/ahr/118.5.1345.

Gregory, D., 2011. The everywhere war. *The Geographical Journal*, 177 (3), 238–250. doi:10.1111/geoj.2011.177.issue-3.

Gritzner, C.F., 2003. Transmogrifying regions. *Journal of Geography*, 102 (5), 226–227. doi:10.1080/00221340308978551.

Güney, A. and Gökcan, F., 2010. The 'Greater Middle East' as a 'modern' geopolitical imagination in American foreign policy. *Geopolitics*, 15 (1), 22–38. doi:10.1080/14650040903420370.

Harris, K., 2016. Making and unmaking of the greater Middle East. *New Left Review*, 2 (101), 5–34.

Harvey, D., 1973. *Social justice and the city*. Athens, GA: University of Georgia Press.

Harvey, D., 2001. Globalization and the 'spatial fix'. *Geographische Revue*, 2, 23–30.

Högbladh, S., 2012. Peace agreements 1975–2011: updating the UCDP peace agreement dataset. *In*: P. Therése and L. Themnér, eds. *States in armed conflict 2011*. Uppsala: Uppsala University Department of Peace and Conflict Research, 39–56.

Huntington, S.P., 1993. The clash of civilizations? *Foreign Affairs*, 72 (3), 22–49. doi:10.2307/20045621.

Khalil, O., 2014. The crossroads of the world: U.S. and British foreign policy doctrines and the construct of the Middle East, 1902–2007. *Diplomatic History*, 38 (2), 299–344. doi:10.1093/dh/dht092.

Khalil, O., 2016. *America's dream palace: Middle East expertise and the rise of the national security state*. Cambridge, MA: Harvard University Press.

Klare, M., 2004. *Blood and oil: the dangers and consequences of America's growing dependency on imported petroleum*. New York, NY: Metropolitan Books.

Koch, N., 2016. Is a "critical" area studies possible? *Environment and Planning D: Society and Space*, 34 (5), 807–814. doi:10.1177/0263775816656524.

Kuper, A., 2005. *The reinvention of primitive society: transformations of a myth*. New York, NY: Routledge.

Kymlicka, W. and Pföstl, E., eds., 2014. *Multiculturalism and minority rights in the Arab world*. New York, NY: Oxford University Press.

Latour, B., 1993. *We have never been modern*. Cambridge, MA: Harvard University Press.

Lawrence, B., 1998. *Shattering the myth: islam beyond violence*. Princeton, NJ: Princeton University Press.

LeFebvre, H., *et al.*, 2003. Space and the state. *In*: N. Brenner, ed. *State/space: a reader*. Malden, MA: Blackwell, 84–100.

LeFebvre, H., 2009. Reflections on the politics of space. *In*: N. Brenner and S. Elden, eds. *State, space, world: selected essays*. Minneapolis, MN: University of Minnesota Press, 167–184.

Lewis, M.W. and Wigen, K., 1997. *The myth of continents: a critique of metageography*. Berkeley, CA: University of California Press.

Lockman, Z., 2004. *Contending visions of the Middle East: the history and politics of Orientalism*. New York, NY: Cambridge University Press.

Lockman, Z., 2016. *Field notes: the making of Middle East studies in the United States*. Stanford, CA: Stanford University Press.

Lustick, I.S., 1997. The absence of Middle Eastern great powers: political backwardness in historical perspective. *International Organization*, 51 (4), 653–683. doi:10.1162/002081897550483.

Lynch, M., 2016. *The new Arab wars: uprisings and anarchy in the Middle East*. New York, NY: PublicAffairs.

Mamdani, M., 2012. *Define and rule: native as political identity*. Cambridge, MA: Harvard University Press.

Mbembé, A., 2003. Necropolitics. *Public Culture*, 15 (1), 11–40. doi:10.1215/08992363-15-1-11.

Messner, J.J., ed., 2018. *Fragile states index*. Washington, DC: The Fund for Peace.

Miliband, R., 1977. *Marxism and politics*. New York, NY: Oxford University Press.

Mitchell, T., 1991. The limits of the state: beyond statist approaches and their critics. *American Political Science Review*, 85 (1), 77–96. doi:10.1017/S0003055400271451.

Mitchell, T., 2002a. McJihad: islam in the U.S. global order. *Social Text*, 20 (4), 1–18. doi:10.1215/01642472-20-4_73-1.

Mitchell, T., 2002b. The Middle East in the past and future of social science. *In*: D. L. Szanton, ed. *The politics of knowledge: area studies and the disciplines*. Berkeley, CA: University of California Press, 74–118.

Mitchell, T., 2011. *Carbon democracy: political power in the age of oil*. New York, NY: Verso.

Møller, F.S., 2017. Blue blood or true blood: why are levels of intrastate armed conflict so low in Middle Eastern monarchies. *Conflict Management and Peace Science*. doi:10.1177/0738894217714716.

Mundy, J., 2011. Deconstructing civil wars: beyond the 'new wars debate. *Security Dialogue*, 42 (3), 279–295. doi:10.1177/0967010611405378.

Nitzan, J. and Bichler, S., 2002. *The global political economy of Israel: from war profits to peace dividends*. New York, NY: Pluto Press.

Olimat, M.S., ed., 2014. *Handbook of Arab women and Arab Spring: challenges and opportunities*. New York, NY: Routledge.

Owens, P., 2015. *Economy of force: counterinsurgency and the historical rise of the social*. New York, NY: Cambridge University Press.

Paasi, A., 2003. Region and place: regional identity in question. *Progress in Human Geography*, 27 (4), 475–485. doi:10.1191/0309132503ph439pr.

Paasi, A., 2011. From region to space part II. *In*: J.A. Agnew and J.S. Duncan, eds. *The Wiley-Blackwell companion to human Geography*. Somerset: John Wiley and Sons, 161–175.

Paasi, A. and Metzger, J., 2017. Foregrounding the region. *Regional Studies*, 51 (1), 19–30. doi:10.1080/00343404.2016.1239818.

Pettersson, T. and Eck, K., 2018. Organized violence, 1989–2017. *Journal of Peace Research*, 55 (4), 535–547. doi:10.1177/0022343318784101.

Poulantzas, N., 1978. *L'État, le pouvoir, le socialisme*. Paris: Presses Universitaires de France.

Rankin, W., 2018. *After the map: cartography, navigation, and the transformation of territory in the twentieth century*. Chicago, IL: University of Chicago Press.

Roberts, H., 2007. *Demilitarizing Algeria*. Washington, DC: Carnegie Endowment for International Peace.

Ross, M.L., 2018. The politics of the resource curse: a review. *In*: C. Lancaster and N. van de Walle, eds *The Oxford handbook of the politics of development*. New York, NY: Oxford University Press, 200–223.

Rowley, R., Bichler, S., and Nitzan, J., 1989. *The armadollar-petrodollar coalition and the Middle East*. Department of Economics Working Papers, McGill University.

Said, E.W., 1978. *Orientalism*. New York, NY: Vintage.

Sarkees, M.R. and Wayman, F., 2010. *Resort to war: 1816–2007*. Washington, DC: CQ Press.

Sayer, D., 1987. *The violence of abstraction: the analytic foundations of historical materialism*. Oxford, UK: Blackwell.

Scahill, J., 2013. *Dirty wars: the world is a battlefield*. New York, NY: Nation Books.

Schrodt, P.A., 2014. Seven deadly sins of contemporary quantitative political analysis. *Journal of Peace Research*, 51, 287–300. doi:10.1177/0022343313499597

Scott, J.C., 1998. *Seeing like a state: how certain schemes to improve the human condition have failed*. New Haven, CT: Yale University Press.

Sidaway, J.D., 1994. Geopolitics, geography, and 'terrorism' in the Middle East. *Environment and Planning D: Society and Space*, 12 (3), 357–372. doi:10.1068/d120357.

Sidaway, J.D., 2013. Geography, globalization, and the problematic of area studies. *Annals of the Association of American Geographers*, 103 (4), 984–1002. doi:10.1080/00045608.2012.660397.

Sørli, M.E., Gleditsch, N.P., and Strand, H., 2005. Why Is there so much conflict in the Middle East? *Journal of Conflict Resolution*, 49 (1), 141–165. doi:10.1177/0022002704270824.

Stampnitzky, L., 2013. *Disciplining terror: how experts invented "terrorism"*. New York, NY: Cambridge University Press.

START (National Consortium for the Study of Terrorism and Responses to Terrorism), 2018. *Global terrorism database*. College Park, MD: University of Maryland.

Stavrianakis, A. and Stern, M., 2018. Militarism and security: dialogue, possibilities and limits. *Security Dialogue*, 49 (1–2), 3–18. doi:10.1177/0967010617748528.

Streeck, W., 2014. *Buying time: the delayed crisis of democratic capitalism*. Translated by P. Camiller and D. Fernbach. New York, NY: Verso.

Tilly, C., 2003. *The politics of collective violence*. New York, NY: Cambridge University Press.

US Arms Control and Disarmament Agency, 1975. *World military expenditures and arms transfers 1963–1973*. Washington, DC: U.S. Arms Control and Disarmament Agency.

US Arms Control and Disarmament Agency, 1982. *World military expenditures and arms transfers 1970–1979*. Washington, DC: U.S. Arms Control and Disarmament Agency.

US Arms Control and Disarmament Agency, 1984. *World military expenditures and arms transfers 1972–1982*. Washington, DC: U.S. Arms Control and Disarmament Agency.

Vitalis, R., 2007. *America's kingdom: mythmaking on the Saudi oil frontier*. Stanford, CA: Stanford University Press.

Vitalis, R., 2015. *White world order, black power politics: the birth of American international relations*. Ithaca, NY.: Cornell University Press.

Vitalis, R. and Heydemann, S., 1999. Explaining state-market relations in the post-war Middle East. *In*: S. Heydemann, ed. *War, institutions, and social change in the Middle East*. Berkeley, CA: University of California Press, 100–145.

Zulaika, J. and Douglass, W., 1996. *Terror and taboo: the follies, fables, and faces of terrorism*. New York, NY: Routledge.

Normative Scaling and Crisis Knowledge: The Problematic Use of Selective Analogies to Compare Conflicts

Florian P. Kühn

ABSTRACT

Comparing conflicts intentionally and consciously ascribes meaning to a political subject. When Afghanistan is linked to Vietnam, or its democratic development to Switzerland, normative scales are introduced to order discursive practice. Discovering meaning and making meaning is what Simone de Beauvoir (1947) views as existential condition of human beings: political acts are measured against existing norms of the acceptable. In conjunction with Mannheim's sociology of knowledge (2015), this article explains the social situatedness of crisis knowledge and the norms enshrined therein. In conflict comparisons, reservoirs of knowledge are being connected and repertoires of action prescribed. Based on these theoretical approaches, the article unpacks the power relations and political intentions at work in these comparisons.

1. Introduction: Discovery of Meaning and Meaning Making in International Interventions

In protracted and complex conflict dynamics, decision-makers orient knowledge following analogies from other conflicts. Comparing one conflict with another serves the purpose of activating normative assumptions. According to those comparisons, this article argues, normative positions are being defined, which hierarchise not only policy priorities but also people subject to policy. When it is argued that Afghanistan is not Switzerland, this comparison evokes a second-class Afghan citizen whose entitlement to democratic procedures and to well-being is somewhat less pronounced than it would be for a Swiss citizen. If Afghanistan is being related to Vietnam, the analogy highlights the protracted fighting and possible domestic political fall-out. These statements underpin knowledge and, in turn, justify policy formulated by outside intervention professionals. The practice of using selective analogies to ascribe meaning to conflicts and resultant interventions is problematized in this article, highlighting the following questions:

who makes a comparison? – *how* is being compared? – along *which* explicit or implicit *categories*? – what is the *purpose* of comparison? – which are the *ethical implications* of categories at work?

These questions help discern interests of discourse participants, patterns of ordering as well as the normative starting points from which meaning-making emerges. On the surface, establishing comparisons is the practice of discovering meaning – arranging already existing aspects of knowledge creates connexions between cases. Comparisons' meaning making, in extension, creates orders of knowledge, and enables certain ways of doing politics towards a subject matter, while excluding others.

Comparing Afghanistan with Vietnam, or highlighting difference to Switzerland, thus on the surface is merely a technique to connect knowledge; but it also, on a deeper level, assigns Afghanistan its place in the world, where it can be objectified and where its social dynamics can be left out of consideration. A speaker's political intention is intuitively understood by an audience based on commonly shared knowledge. For example, comparing the United States with Pakistan (de Lombaerde *et al.* 2010, 741), or other seemingly quite different cases, would require first formulating a framework which makes a comparison 'possible'. Why common understanding can be accessed when comparing Afghanistan and Switzerland, or Afghanistan and Vietnam, is what gives us clues about the political intentions behind that comparison and reveals the normative scaling at work.

This article combines Simone de Beauvoir's insights on the conquest of reality by man with Karl Mannheim's sociology of knowledge as socially bounded, not absolute, knowledge. This combination provides novel insights into how practice in international politics relies on certain types of knowledge, while devaluing others. Claiming superior expertise and insights, as de Beauvoir argues, guides actions which in turn create new insights to frame and understand an environment. Mannheim's concept allows unpacking the conditions under which knowledge becomes salient and accepted. De Beauvoir highlights the will to know and the will to mould reality, and shows the structural preconditions under which this takes place. From these theoretical grounds the article analyses the scaling efforts to ascribe meaning to the war(s) in Afghanistan. Scaling, I argue, establishes normative boundaries that facilitate a knowledge group's powerful social position to be legitimised. Intervention professionals who share a distinctive interest as expatriates and conflict experts, use analogies of Afghanistan to establish knowledge useful to themselves. Portraying Afghanistan in a version that makes it administratively and normatively manageable has become one of the tasks of these intervention professionals to sustain their influence and political position. The article concludes by tying these findings back to their theoretical starting point and sketching ways for future research.

2. Interventions and Normative Scaling

Interventions are born from an international system putting the national state with a central government at its centre. State institutions are viewed as main arena of political conflict, disregarding other forms of conflict management or, indeed, politics outside state structures (Lemay-Hébert 2009, 23). In Afghanistan, the Taliban, despite limited success at centralising government, have been complicit in reifying the national scale as the dominant level of politics despite social practice being much more fragmented and localised (Dorronsoro 2005, 339; Giustozzi 2008, 102–113; 2009, 300, Shahzad 2011). Kühn (2011) has argued that the trajectories of state formation impact on societal structures in ways usually not recognised by outside actors. For Afghanistan, the assumption of post-colonial society formation dominates despite fierce epistemic resistance by Afghans against reading 'their' history through such a lens; Afghanistan's political economy is shaped by rentier income (see Rubin 2002, 64–66) rather than taxation or, as would be the case in a colonial setting, funnelling through of extractive wealth (Kühn 2008; Suhrke 2011).

Such functional 'cornerstones' of statehood provide the categories along which normative scaling creates knowledge about a place. It actualises knowledge transferring impressions from other settings. Helped by debates in the 1990s about human security, in the 2000s by the conditions for sovereign statehood defined through the 'responsibility to protect' framework, and recently in resilience debates, these scales measure Afghanistan against an abstract set of state tasks. As Rist argued for development more broadly, the historical conditions that would explain the 'lead' of some countries over others cannot enter into the argument, since the 'laws of development' are supposedly the same for all (...) as if the existence of industrial countries did not radically alter the context in which candidates for industrialisation have to operate' (2002, 74–75).

Consequently, countries which do not reach respectable marks for development and human security, and who fail to develop institutionalised conflict-management and to guarantee citizens' rights, face less favourable terms of engagements (including economic terms of trade); they are less likely to be integrated into formal economic networks of exchange. In security policy, reaching 'par' level by not threatening the international system through fragility, ungoverned spaces, and toleration of sub-state violent actors, is prerequisite to international recognition (Bueger and Gadinger 2018, 109, Boltanski 2011, 54–55, Chandler 2010; Pouligny 2006). In self-stabilising cycles of mutually interlocking arguments – economic, legal, humanitarian, levels of 'development' – states are scaled and, consequently, treated as mature or inferior and hierarchically subordinate (Kühn 2010, 148–171). Comparisons establish such scales in that they give orientation by levelling a given state; in comparison, another country may fulfil such standards, or fall short of acceptable levels of stateness. A minimal level on such a normative scale, such as adhering to international standards of human rights, or diplomatic practice, would be required for a state to escape political and normative trusteeship (Hirblinger and Landau 2018, 390, Hameiri *et al.* 2018, 470).

Establishing normative scales, of course, depends on rhetorical tactics of the intervention professionals who continually evaluate and re-evaluate politics in relation to its own practice, producing conflict knowledge. Knowledge production has also become an explicit part of international organisations' activities (Bliesemann de Guevara and Kostić 2017). Knowledge is rhetorically embedded in professionalism, as Goetze illustrates: 'The elusive nature of the notion of "professionalism" in peacebuilding, and its weak relation to specific skills, knowledge, or even standardised protocols, emphasises how much the discourse of professionalism is part of self-description and self-identification. ... Collectively, the discourse of professionalism draws boundaries between different organisations as it allows hierarchization along lines of professional efficiency, whether or not this professionalism is anchored in real, measurable entities' (Goetze 2017, 164). Hierarchies within the group of professionals in interventions are established around forms of knowledge. In turn, knowledge needs to be made accessible to those outside the group, i.e., media, politicians, and public. Credibility is, to adopt Goetze's Bourdieusian terminology, convertible into power and political influence, defining problems and delivering solutions.

While the sociology of interventions is a valuable yet understudied field of enquiry, the ontology and epistemology of how conflicts are debated, judged, and eventually understood, is even less well understood. Heathershaw (2016) describes the 'international community' as group of internationals with comparable backgrounds of specified academic programmes, with a relatively homogenous cosmopolitan perspective and roughly shared norms (see Bliesemann de Guevara and Kühn 2011; Goetze 2017, 95–100). Establishing survey marks by introducing comparisons thus represents a process of pinning down norms which can be used by conflict professionals to make sense of a situation in which they are politically engaged – as well as create a self-identification. As Hirblinger and Landau (2018, 391; see also Hameiri and Scarpello 2018) point out, and as is theoretically developed by Mannheim (2015 [1929], see below), this helps create, and stabilise, the epistemic community whose entitlement to power is enshrined in the knowledge it produces. Needless to say that earlier experience of these professionals feeds into the base knowledge, with concepts travelling from Bosnia to East Timor, to Banda Aceh and Afghanistan, to Iraq, and Mali.

The distinctive interest of holding on to epistemic power is what inspires the use of the concept of an 'interventionist class'. It derives from Elsenhans' description of (then) 'Third World' state bureaucracies which are centralised and able to allocate state funds to their own benefit while controlling decision-making positions. He calls the group of politicians and bureaucrats the 'state class' (see also Cox 1981, 151). The analogy finds its limits in national narratives and habitus, but works for the 'self-privileging' of functional and political elites (Elsenhans 1981, 24–26). It can be observed in international interventions, where individuals may be employed with distinct duties as diplomats or humanitarian aid workers by national governments and their agencies, as members and employees of international organisations such as the UN, EU or World Bank, as non-governmental and/or philanthropist activists[1]. Like members of a 'state class', intervention professionals differ in political outlook and policy orientation, but undergo socialisation through various 'rites of passage', thereby creating a cohesive social group that occupies a liminal space (Smirl 2015, 42–45, Heathershaw 2016; Goetze 2017 on inner fragmentation and stratification). Many of this group – particularly in security-strained environments such as Afghanistan – interact within a social and epistemic 'bubble' (Andersson and Weigand (2015, 524) refer to Afghanistan's capital as 'Kabubble'). A sub-group of the epistemic community, which creates knowledge and, through narratives and social exchange, shares experiences (Bliesemann de Guevara and Kühn 2015) is the small number who might be called 'translators'; they transmit these group-specific narratives into accessible sound bites which are understood by an out-group audience, producing 'legitimacy through shared and oppositional stories. This is because stories are far more adaptive to different policy positions and audiences than mere interests' (Gadinger *et al.* 2019, 93, Freistein and Gadinger forthcoming).

Part of broader narratives, comparisons made by these translators addressing a general audience and non-specialist policy makers are functional for the politics of the interventionist class. The question 'who compares' focusses on the network of transnational conflict professionals who self-identify and draw distinctions between them and their political subject. The categorisations follow level-of-statehood criteria. The following two sections will address the question about the purpose of comparison by connecting de Beauvoir's philosophy of how man creates surrounding reality by ascribing meaning with Mannheim's sociology of group-inherent knowledge and how it shapes power-relations.

3. 'Serious World': The Acceptable and the Stabilisation of the Desirable

What Simone de Beauvoir calls the 'serious world' is the tension an individual spirit experiences when realising the world as subject to her agency. At the same time, she realises she is subject to and shaped by external conditions and forces. On one hand, 'the characteristic of the spirit of seriousness is to consider values as ready-made things' (de Beauvoir 1948, 35). Soon, on the other, values and rules are questioned in order to discover their meaning, and their validity is challenged once their meaning is approached. Intending to 'manage' meanings (disclosing, making, or simply the desire to ascribe these), human beings cannot stop engaging with the meaning of themselves as subjects and of social circumstances. De Beauvoir describes this state as 'ambiguous' and indefinite, since neither the desire to discover the meaning of the world, nor the desire to make the meaning of the world in a godlike fashion can be fulfiled. Discovering and creating meaning, and the knowledge which underpins it, is essential to individual existence in social relations.

Two forms of intentional human activity can be discerned: first, revealing the meaning which, apparently, exists independently of oneself; and second, the active and directed striving for the creation of meaning. In the first, consciousness expresses its freedom to discover meaning. In the second, it uses its freedom to become the author of the meaning of the world (Bergoffen 2015). Since this ambiguous state persists, an ethical position needs to be developed in place of otherwise inevitable nihilism due to humans' inability to reach fulfilment. Any ethics deriving from the ambiguous existence cannot rely on absolutes, such as God or truths beyond human comprehension (also Bauman 1992, 335–348)

This notion of meaning can inform analyses of conflict, and by extension, of conflicts comparison. De Beauvoir's ambiguous function of the freedom of the mind operates in international conflict settings. The discovery of meaning, ostensibly existing 'somewhere out there', is a central aspect of the work of conflict professionals: Analysts, policy makers, and journalists, but also perpetrators of violence constantly attempt to make sense of what they see as the essence of a conflict. This essence is permanently re-evaluated, and negotiated between several actors as part of their drive to unearth the conflict's core meaning. De Beauvoir points out the temporality of these negotiations, as claiming absolute positions means sacrificing the present for the future: conflict over meaning transcends the present and turns it into something instrumental to achieve what is only to become in the future. In other words, the conflicts of today are fought over the state the world ought to be in in the future.

While conflicts and violence shift shapes and change their core content as they evolve (Elwert *et al.* 1999, McAdam *et al.* 2001, 72–78, Körppen *et al.*

2011), a static understanding of conflict continues to dominate liberal understandings of conflict. Policy making predominantly strives to address a conflict's 'root causes'. The ambiguous state allows to dissect the surroundings within which meaning is created and enacted. The will to create meaning is enshrined in social institutions, such as law, modes of reproduction, in short: the state. De Beauvoir argues

> "... if it is true that men seek in the future a guarantee of their success, a negation of their failures, it is true that they also feel the need of denying the indefinite flight of time and of holding their present between their hands. Existence must be asserted in the present if one does not want all life to be defined as an escape towards nothingness." (1948, 125).

Cassirer's (1946) powerful description of the state as its own myth-maker underscores the importance of state institutions; the state institutionally creates history and knowledge to justify not only its existence but also the very form a state takes in historic contingency. The state myth is not necessarily a foundational one but explains how the state has become exactly what it was meant to be. Acting groups within institutions of the state are self-empowering – and understand their own position as a necessity (Cassirer 1979, 202, Kühn 2016b, 149–153). Contrasting de Beauvoir's ethics, Cassirer views the systematic disregard for institutions' historicity as a rejection of responsibility for the wrongs these inevitably produce. The interventionist class, in this reading reifies its own existence by ascribing itself the value of a state institution – which it is undoubtedly, given that it is an outgrow of stability desires of the international state system. As, inevitably, the human mind and human institutions cannot remain focused on the present, but 'must be immediately engaged in a new undertaking, ... dash off toward the future' (de Beauvoir 1948, 126), this institution's durability becomes part of its purpose. Full of the past, state institutions thus busy themselves creating the conditions for current and future policy.

The 'serious [wo]man', deeply imported with meaning she finds pre-existing to her own position in the world, acts within institutions to make meaning as the second part of the ambiguous endeavour which is life. Mannheim's sociological approach to knowledge is helpful in unpacking how this meaning is created, and how international organisations engaged in international interventions practice the meaning-making activity of the ambiguous state. Mannheim argues that the frame of knowledge is set by circumstance. Knowledge is not only influenced, but 'decisively determined in content and form, in matter and in the way of formulation, capacity, intensity of the grasp of the bearings of experience or observation, in one word everything which we will call the structure of the aspects of insights' (Mannheim 2015 [1929], 230; author's translation).

In the next section, I will show how the idea of a social perspective enshrined in knowledge complements de Beauvoir's distinction between approaching vs creating meaning. In order to prepare the ground for a conceptual inroad towards conflict comparison, normative scaling connects these concepts to derive insights from Afghanistan.

4. Mannheim's Perspectivity of Knowing

Mannheim (2015 [1929]) shows knowledge to be contingent upon the social circumstances of its becoming. He explains how knowledge is not only influenced, but systematically depends on and is consequently shaped by the social perspective on a problem about which knowledge is acquired; he strictly avoids a metaphysical grounding of epistemology (2015 [1929], 267; 1980). His is not an approach to de-mask ideologies as necessarily false beliefs, but to reconstruct the conditions of how knowledge is produced (see Cox 1981). The 'structure of aspects of insights' (Aspektstruktur) directly leads us to the existence and the positionality of the person having such insights (2015 [1929], 229). A sociology of knowledge, thus, reconstructs the material, epistemic, historical conditions under which knowledge comes to gain relevance. It is unimportant whether an ideology is purposefully deceitful or is believed by those who deem it based on knowledge. Instead, what is important is the question of how knowledge relates to its subject matter, how it operates, and how the politics of knowledge shape what counts as knowledge in the first place.

Mannheim insists that knowledge is relational and needs to be viewed within the bigger picture of its conditions of existence. Perspectivising it ('Perspektivierung') means to connect different types of knowledge to different types of groups who produce knowledge which serves as their guiding practice. Such leading groups – not necessarily classes, but also 'generations, peers, sects, professions, schools' (Mannheim 2015 [1929], 237) define the ontology of a subject. The more powerful a group is relative to other competing centres of power, the more dominant its specific group knowledge is to be assumed. Gramsci's (1991) 'hegemony' assumes those subject to a system of rule deem this system justified – such justification, naturally, rests on a certain type of knowledge underpinning it.

Connecting de Beauvoir's two types of dealing with meaning – discovering it, creating it – to Mannheim's insights shows that the international interventionist class carries the ambiguous process of ascribing meaning to the politics of international interventions, while simultaneously creating that world by defining its political parameters. De Beauvoir points us to the near tragic dilemma of social action of shaping an environment while at the same time trying to make sense of it. Mannheim amends this, arguing that the creation of knowledge is key to (self-)establishing actorness in the first place. A group able to define a subject is able to devise its own legitimacy; it creates its own social position as well as institutional and legal hierarchies in the process (Sending

2015). In international interventions, we observe that the elites guiding the policy of interventions (i.e., intervention professionals) are the same ones who get to define the problem. Consequently, a problem requiring interventions will undergo reframing when those professionals need to adjust policy in the face of limited success.

The following section describes the interventionist class and how it shapes the knowledge of international discourse. In making the meaning of Afghanistan, the interventionist class fabricates the conditions of its room for political manoeuvre, e.g., defines what is say-able about and what is do-able in policy. Comparing aspects of the conflict is one of the mechanisms, which has been introduced as 'normative scaling'. It is a discursive tactic within the confines of this group's knowledge to shape the subject matter. The next section illustrates how comparing Afghanistan helped lower the bar of what was expected from the intervention. I focus on the period of mounting intervention fatigue on the side of interveners (roughly 2008–2014), which illustrates how a notion of 'good enough governance' (Grindle 2004, Hönke 2013, 23–24) was established as a minimum threshold for what could be achieved by outside actors in Afghanistan.

5. Comparing Afghanistan to Switzerland, and Other Places

Afghanistan specialist Ruttig (2017), in his blog collected some comparisons of Afghanistan with other countries. The comparison establishes a scale of violence and social order against which the situation in Afghanistan is being judged. This helps justifying distinct policies, supported by the respective speakers. For example, Ryan Crocker, then US ambassador to Kabul, declared that the biggest risks in Kabul stem from traffic instead of the Taliban (Borger 2011), in order to highlight that the intervention was 'winning' the war. Franz-Michael Mellbin, EU-ambassador to Afghanistan, found that 'The security situation varies from place to place in Afghanistan. You cannot go everywhere, but you can certainly go to a lot of places, especially the bigger cities, and of course the capital Kabul, which has attracted a huge number of people, including the internally-displaced persons' (Deutsche Welle 2016); thus, he supported the right-wing-induced demands to return refugees to countries of origin which could legally only been done without subjecting them to intolerable danger.[2]

These comparisons and statements in support for a domestic agenda are aimed at an audience outside of Afghanistan; they are a constituent part of what Daxner has called 'homeland discourse' (Heimatdiskurs) – a set of knowledge, 'collective taking notice of a certain development' (Daxner 2012: 25), filtered by structures of power and knowledge. This discourse 'produces' a problem, requiring a solution which is implied in how the problem is framed. Comparing the situation in a country understood to be in crisis, is then an act of translation into language and knowledge fuelling a discourse whose participants know little about the actual situation. The normative scales of the comparison tie what one wants to say about a foreign situation to what most discourse participants understand. Doing this, a speaker necessarily will select, and reduce, what she is declaring relevant to find acceptance within an audience. In other words: creating a common ground of understanding a problem is motivating such comparisons. As Read (2019) points out in this Special Issue, statistically scaling a problem works to internationally highlight what ought to be on the agenda; parallel, normatively collocating problems underscores, or devalues, alternative policies. De Beauvoir's 'will to discover meaning' goes hand in hand with creating it, as the interventionist class engages in defining a reality which becomes the backdrop of its own action. Between policy makers' sloppy way of comparing and a situation on the ground stand the professionals who provide the knowledge in which such influential utterances are (sometimes more, sometimes less) grounded.

The ostensible unruly chaos of Afghanistan does not fruitfully compare to the assumed orderliness and structured social relations of Switzerland. Ascribing meaning to such contrasting trajectories centrally fulfils de Beauvoir's understanding of meaning-making. It is between the intervention's protagonists – diplomats, UN personnel, aid agencies – where this comparison was observed most prominently. Comparing Afghanistan to Switzerland goes back a long way, but in surprising connexions. According to Maley (2013), 'In 1937, the French diplomat René Dollot described Afghanistan as "the Switzerland of Asia". Afghanistan was a long way from Switzerland, but in a Europe haunted by the rise of fascism, this remoteness gave it a certain appeal'. The contrast underscored by the comparison is the illocutionary motivation of juxtaposing these countries. 80 years ago, the remoteness Dollot alludes to marked the exception to European countries wedged together in war, extremism and confrontational politics. It remains a speculation if the comparison was made merely based on mountainous scenery, or to exotise Switzerland within the European landscape. Today, comparing Afghanistan to Switzerland evokes romanticised images of peaceful cows on mountain slopes, cheese and the occasional bank, all in sharp contrast to the unpeaceful, unorderly images of Afghanistan. In a vivid illustration of how misleading statistics can be, Switzerland's civilian arms possession is at 27.6 per 100 citizens rated more than double that of Afghanistan (12,5/100; Small Arms Survey 2018). The Bonn International Centre for Conversion's Global Militarization Index (GMI) ranks Switzerland on 40th, and Afghanistan 75th, with amazingly similar scores in the Military Personal and Military Expenditure indices (BICC GMI 2017). On a political systems scale between consensual and majoritarian politics (Lijphart 1999), both would be rather far apart.

Ruttig (2010) argues that, institutionally, the differences between cantonal Switzerland and centralised Afghanistan could not be bigger. This leads him to believe that such comparisons are deliberately omitting some facts.[3] Even if this is the case it may still be legitimate to compare Switzerland and Afghanistan, but making such a comparison itself provokes questions about the political intention behind it. Thus Ruttig (2010) dissects how the Western-dominated 'international' community' (Bliesemann de Guevara and Kühn 2011) used this comparison to prepare a way out of Afghan politics. The 2008 Afghan presidential election marked a decisive policy adjustment (Maley 2009, 270, Gentilini 2013, 112–3). International indifference about observing, which in fact meant far-reaching tolerance of electoral fraud, represented a distancing from the second Karzai government. Corrupt practices from the beginning and actively changing the political landscape led the interventionist class to reject indirect 'ownership' of the dysfunctionalities of the Afghan political system much earlier though. In the bigger picture, this underscored the delegitimisation of central state politics for a large majority of Afghans during the agony years of the Karzai government (Chandrasekaran 2012, 228–240).

The project for a political revival of Afghanistan, creating institutions through which citizens ought to engage and influence policy making after 2001, was coming to an end, while statebuilding to stabilise continued. The comparison, in this context, represents the lacklustre recognition of a socio-political order defined by local violent power relations and alternative systems of governance not in line with the external intervention's objectives. While Western academia is indulging in post-colonialism jargon, the continued experience of structures of oppression very much remains a fact of life for many (Calvert and Calvert 2014, 18–19; also Turner in this issue). International hegemonic structures remain at work in UN mandating-practice and debates as well as in World Bank routines, e.g., the software of capitalist interventions. Statebuilding ideas of the 1990s with their hubris of attempting to re-create from scratch societies and their administrative structures from a *deus-ex-machina* position (Paris 2004) have, with the 2010s, lost most of its salience[4].

The failure of broader Western-led statebuilding ideas is thus revealed, when with a narrower focus Afghans are blamed for not starting to behave like the Swiss ostensibly would have: orderly, assiduously, with fine-tuned compromise rebuilding the country: 'Western leaders seeking to exit from an unhappy theatre of operations have readily slipped into a discourse in which existential flaws in Afghanistan and Afghans – rather than Pakistani meddling or the attempt to establish a strongly centralised state to rule over a kaleidoscopic collection of micro-societies – carry the blame for what has gone wrong' (Maley 2013, 215). Deflecting such responsibility by pointing to its conceptual impossibility is the message these comparisons carry:

> "These elections – 'at the worst possible time', as UN envoy Stefan de Mistura just said on [sic!] the BBC – is part of a larger exercise called 'Afghanisation' and 'transfer of responsibility' to the Afghans. Who, by the way, are supposed to be in the lead already [sic!]. In principle, transferring responsibility makes a lot of sense. Of course, the Afghans need to be in charge of their own affairs. But institutions are still lacking through which they can develop a genuine national consensus on major political issues. In contrast, the Taleban line about a foreign occupation and a necessary fight for independence already resonates with people who otherwise abhor them" (Ruttig 2010, n.p.).

6. What It Means to Compare

Obviously, there is a gap between the knowledge and the sloppy everyday, newspaper-headline comparison.[5] Notably, this type of comparative discourse works entirely without input from persons about whose life-world such comparisons and derived policy scales are being made (Hameiri *et al.* 2018, 469). By extension, decisions taken in Brussels, New York, Washington, London, Ottawa, or Berlin, will likely be more informed by a detached flow of (public) opinion and information than by first-hand experience. Regrettably or not, this is not only the mode of discussions about crises (Schneckener 2005). Expertise guides policy making in domestic politics of all topics and challenges: In institutionally diverse ways, states and other forms of social organisations channel expert input and the view of affected parties. Senate and House committee hearings or the appointment of academics to the Privy Council Office, extra-parliamentary ad-hoc advisory councils, or trips by parliamentarians to 'see for themselves' are ways of grounding and creating credibility and, consequently, legitimacy for decisions, notably when these are about matters war and peace (Bliesemann de Guevara, 2017). In turn, being heard testifies to the policy relevance and 'impact' of academic researchers.

Yet, comparison creates categories which in turn create orders of values. The normative scales which facilitate intersubjective understanding are providing orientation. Comparing, for example, Afghanistan as an intractable war-zone with Vietnam, presupposes a connexion between core elements of ulterior discourses – understanding both as a war which cannot be won due to unbreakable resistance of the powerless, or alternatively, as struggle of the powerless against (neo-)colonial suppression and crude force. It also connotates domestic political fall-out of disenfranchised youth, peace movement and delegitimisation of military action. Either version, in an ambiguous sense, is true – the discursive in-group building strategy, however, aims at creating a normatively bound epistemic community (Kühn 2016a). Talk about 'Islamo-fascism', or comparing the so-called War on Terror to the systemic clash of the Cold-War confrontation between communism and liberalism, was the back-drop of and guiding the decision to invade Afghanistan after 2001 (Dorronsoro 2005, 318–9). Building categories thus builds in-group understandings, by presenting a reading as 'established' and hence universally 'valid' and 'accepted'. By excluding aspects of the bigger picture, this strategy closes off debate about different, equally valid readings (Maley 2013). Categories are scaled to claims of policy relevance, urgency and the set of available policy options which are ingrained in how they are conceptualised. It is thus not only a 'common sense' reaction to reality, but invested with resources to understand, and create meaning.

Furthermore, comparison of conflict and violence systematically (and one might speculate: deliberately) suppresses details of violence. As in Tolstoy's *Anna Karenina principle* '[h]appy families are all alike; every unhappy family is unhappy in its own way', wars are always dissimilar in the ways violence creates its own preconditions during processes of violent escalation. In other words, how wars are fought and mass violence brought to bear in support of a political goal (excluding amok, domestic, or spontaneous mob outbreaks of violence) create path-dependent structural conditions shaping the violence (von Trotha 1997, 32–33). The structures, ways of violence, and actors within a set of dynamic interactions in war can be compared, but at the cost of neglecting the complexity of violence and how it unfolds in a concrete case. Mostly however, statistical analyses are limited to simply comparing conflict(s) along selected operating figures (Cramer 2006, 57–86, Kühn 2016b, 155–156). That way, naturalising claims are being made which essentialise the 'true' core of a conflict, to be addressed by extracting the 'root' causes.

Finally, comparison connects conflicts across time. Connecting across some decades, but even a few years, forces *zeitgeist*-reading onto the interpretation of an older conflict which was understood differently by contemporaries. Implicit knowledge based on evolving interpretations of discourse participants implicitly informs such references. If, for example, Afghanistan is being compared to Vietnam, the backdrop of the Cold War is largely lost today (along with readings of resistance against communist expansion in the 'Third World', domino theory, or of the continuation of a decolonising struggle). With the benefit of some hindsight, Ruttig's (2010) analysis was chillingly foresighted. The political purpose of the Switzerland-Afghanistan comparison aims to cover up intervention fatigue by declaring a natural (and thus essentializing) Afghan propensity for chaos and unruly (political) behaviour dating back 'centuries' (see Kühn 2016a). Ruttig points out with some frustration that the political intention of delegitimizing Afghan politics is aided by the analytical meaninglessness of the comparison.

Summarising, it has been established that comparison works without first hand input, knowledge or including people's interests. It rather establishes normative hierarchies which allow positing policy priorities exogenic to the political 'problem' at hand. Counter-intuitively, violence within such comparisons plays a minor role; it works as a token to depict something as problematic, or by downplaying responsibility for it, but remains opaque in that its dynamic hardly enters the political picture. Across time, thus, the comparisons become analytically meaningless while this void of actual content allows them to be employed just the more effectively in political discourse.

7. Discovering Meaning, Creating Meaning: Notions of Afghan State Formation

Following de Beauvoir, the meaning making function of international interventions, accompanied by the discourse about them, help unpack the 'serious world' of international politics. Where political engagement takes place, a framing narrative will not only try and make sense of the world, but will aim at shaping it. De Beauvoir points out that meaning-making is a power-laden and, not only in interventions, violent undertaking. The social situatedness of knowledge, according to Mannheim, prescribes a frame within which knowledge about a political endeavour is created, upheld, reproduced, and influences policy. A Western interventionist class, based on the epistemic software and shared points of reference at its collective disposal, uses comparisons in an ordering fashion. Imposing such order through the state and its institutions, along with the law enshrined in it, social difference is acknowledged, even welcomed, but not part of the knowledge and kept outside the statebuilding endeavour.

If state-formation historically was a contingent process, unfolding under conditions of an emerging state system, the functional role of foundationalist reasoning becomes apparent. Only the notion of permanence can insinuate that statehood is the 'mode to be', the Westphalian order a 'convention' to aspire to (Krasner 1994; Jung 2001, 448). For the interventionist class, the concrete form of the state is subsumed into a template model of statehood. In Afghanistan, the modern state was created mimicking emerging states in Europe, themselves product and producer of colonialist expansion and exploitation. Comparing the mode of social organisation to other aspiring states was what influenced the centralisation of state institutions while the social order remained fragmented. Although Afghanistan was never fully colonised, colonialism nevertheless came into and shaped its existence as a state (Barfield 2012). The role of Ahmad Shah Durrani, who united clans and tribes to form Afghanistan and become 'King' (!) was moulded by the model of kings of the time, even attempting to introduce dynastic succession. In late absolutism, the centralisation of violence, of powers to define the law and to control finances were central to defining what the state was to become (Saikal 2004). The Durrani state was, in essence, a warrior state, run by and dependent on military expansion for resources and on military success for legitimacy (Johnson 2011, 40–42). Durrani rulers were trying to adjust to the emerging international order, by comparing their own form of rule, changing what they saw as requirement of stateness in the hope of solidifying their place.

While European powers faced conflicts, and even wars, of succession in spite of rules regulating succession already Ahmad Shah Durrani's succession following his death in 1772 nearly broke the institutions: '... fratricidal warfare between his heirs broke out around the country' (Johnson 2011, 39). Later hand-overs of power were equally violent. In this weakened environment marked by constant threats of in-fighting, a durable centralised institution never consolidated but continually depended on personal loyalty. Balancing tribes, from which different military units had to be co-opted, profiteers from levies on trade and passage, and clerics' demands for true belief, while having to bear in mind external threats to territory and resources, proved too much for most rulers in Afghanistan. For tribesmen and clans, freedom under state rule (in what would be a liberal notion of freedom being possible only under statehood) was less acceptable than freedom without a state at all. The ideal state would, simply, not bother different groups too much with taxes or social regulation.

A good century later, Abdur Rahman 'failed' in similar ways when British interests in the country triggered a second British military intervention (1878–81; after the first Anglo-Afghan War 1839–42). British influence unfolded as part of the struggle with Russia over colonial control and spheres of influence, called the 'Great Game': 'To the Afghan people, Abdur Rahman always argued that his army was organised purely for the defence of Afghanistan against foreign invasion, but in reality the Amir's troops were designed for internal security. Over forty separate disturbances were crushed by the army between 1881 and 1901' (Johnson 2011, 138). Again, comparison with Western states' internal consolidation based on violence creates meaning by establishing scales to measure state formation against: Analogies with European states' enshrined hierarchies would disregard the complexity of societal interaction and negotiations of legitimacy at the time as well as of the question who ought to pay shares in financing and equipping institutions in Afghanistan (Dorronsoro 2005, 28–29).

Ahmad Shah Durrani in the 18th, and Abdur Rahman in the 19th century tried to emulate (modern) statehood in Afghanistan. Colonial powers, notably Britain and Tzarist Russia sought to establish authoritative isomorphic partner institutions in Afghanistan with whom they would be able to conduct foreign policy. Finding 'counter-parts aimed at creating stability, e.g., finding those who would stay in power for a longer period and not be toppled by revolt; it also meant to establish durable relations which might be exploited by bribery, giving and taking favours, and deducing legitimacy claims based on international recognition and networks' (Barfield 2012,

341–2). The 'serious man' of international politics, the state, created meaning through the institutions functional for its own preference within the state system, forcing non-Western rulers to emulate.

It was argued that comparison, instead of merely using existing categories and connecting them in ways akin to the 'technology' of ordering (Foucault 2004, 162–3), creates categories itself: the connexions establish new categories in which relations of super- and subordination are enshrined, of normative scales and of repertoires of policy action deemed appropriate. The 'appropriateness', however, is a result of the categories used for comparisons: when a state is described as fragile, as incapable of governing, as risky (Woodward 2017, Kühn 2017), it becomes part of the serious world which is to be re-created in a less threatening, easier-to-manage manner.

The problem of 'fragile' and 'failed' states to this day consists of a knowledge gap. The interventionist class of diplomats, administrators, aid industry as well as security personnel, have the knowledge and means which the fragile state ostensibly lacks. A unitary state which gives a name and address to those representing the whole of Afghanistan in world affairs, including repaying debt and development loans, requires the knowledge to run a state according to the enforced rules. Afghan leaders on the other hand often only superficially represented but did not speak for their country. Being able to mobilise tribal warriors was an asset in internal competition but did not amount to domination of the population across the board. Effectively, the hierarchy of clans was more important to balance influence on a ruler than military hierarchy based on merits and ability. As a result, changing loyalties would in times of crises become more important than functional roles within the state or its institutions. Mitigating these relations largely depended on external rentier payments instead of a comprehensive taxation regime (Barfield 2012, 153; Rubin 2002).

Tilly's model of war-fighting, e.g., maintaining a military organisation above a certain size, leading to administrative capacity and taxation, is misguided for Afghanistan.[6] Paid largely by the British to keep the Russians in cheque, state-making in Tilly's model was largely missing, or characterised by tribal considerations more than by functional ones (Tilly 1985). Rather, rulers would keep parts of the state weak and incapable in order to prevent contenders from acquiring overwhelming force and military ability (Migdal 1988, 1994). The contemporary interventionist class, with its consultants, trainers, and supervisors, has the knowledge to run a modern state but lacks it about the social exchanges in Afghan politics; trying to discover meaning, consequently, the interventionist class can only spot deficiencies. Their power to enforce meaning, as is the self-ascribed task of the serious man, finds its limits where normative scales do not accurately fit existing social institutions. The logic of latter phases of state formation in Europe, which saw increasing public control of violence after its monopolisation within the state, is misleading in the Afghan context: 'Parallel to the monopolies of physical force and taxation, the modern state also acquired a monopoly over the means of symbolic reproduction. In this process, however, the establishment of legal authority was itself communicated through traditional symbols' (Jung 2001, 460).

As social interconnections develop and are transcending borders, over time (unevenly) integrating political, social and economic reproduction, late modernity is defined by the 'contemporaneity of the non-contemporaneous' (Bloch 1935, quoted in Jung 2001, 456). Comparing conflicts, consequently, rests on emblems of modern statehood as measuring-rods for adaptations not fully implemented. Emblems of the state in Afghanistan may mean different things to the population, but also to Afghan elites, than to the interventionist class. To meaningfully use the analytical potential of academic comparison, the social dynamics of violent and political actions, the economic, social, and political aspects of conflictual transformation need to be situated.

8. Conclusion: De Beauvoir, Mannheim, and the Conflictual Construction of Meaning

In order to make sense of the meaning-making and ordering function of comparison in international politics, the ways in which comparisons are made shows parameters which can be used to analyse their meaning-establishing function for certain social groups: *who* makes a comparison? – *how* is being compared? – along *which* explicit or implicit *categories*? – what is the *purpose* of comparison? – which are the *ethical implications* of categories at work? These questions allows discerning interests of discourse participants, patterns of ordering as well as the normative starting points from which meaning-making emerges. An overarching question remains how we can try to analyse comparisons for their role in the creation of knowledge, and consequently in the creation of meaning? On the surface, establishing comparisons is the practice of discovering meaning – arranging different, existing knowledge discovers the connexions that exist between cases. Much deeper, however, the meaning making, the creation of orders of knowledge is what enables certain ways of doing politics towards a subject matter.

Comparing Afghanistan with Vietnam, thus could solely be analysed as a technique to highlight similarities. It also is, on a deeper level, a way of assigning Afghanistan its meaning as menace to great power politics, according to which it can be objectified and its social dynamics ignored. The striking part is the intuition along which the comparisons are instrumentalised for a speaker's political intention. If comparisons of Afghanistan and Switzerland, or of Afghanistan and Vietnam for example, and of conflicts in general are to be analysed for their political implications, the *speaker's* aims in its discursive context needs to be unearthed in a sequenced approach to describing the embedding conflict dynamics. The discourse is based on a shared field of common understanding between speaker and audience. The commonality is often what Blumenberg (2006) described as myths[7], the power of which is to ground knowledge that needs not be questioned. Tying in historical 'facts' (or rather accepted interpretations of historic narratives) with nodes of meaning to be created stabilises politically charged truth-claims (Kühn 2016b). Myth, thus, underpins knowledge by stating basic assumptions which are widely accepted as true. Myths express certain knowledge about the world, which is untouchable; based on this knowledge, people act and build on it by connecting order through comparison. In social practice, for example of the interventionist class, myths connect to other epistemic 'knowledge', forming a network of mutually supporting tropes – to the extent that reasonable knowledge and presupposed acceptance of something as 'true' becomes impossible to discern (Kühn 2016b, 150).

Essentially, comparing conflicts, countries, societies is a conflictual endeavour itself. Whether the evolution of the Afghanistan intervention is compared to the Vietnam war of the 1960s/70s, or the electoral situation to Switzerland in 2009, making a claim about any conflictual situation is to become a part of the conflict oneself. Thus, the activity to create meaning is ethically charged as argues de Beauvoir. The ambiguity of finding and disclosing the meaning of the world, and attempting to create meaning itself can only be reconciled by the use of epistemic powers: to make others believe the truth one creates by describing the world in a fitting way. Comparing conflicts serves the purpose of establishing a relevant and guiding understanding across different audiences. Using illustrations derived from discourses in and about Afghanistan, I have suggested that not only an academic critique of comparisons is required; also, more importantly, interrogating the circumstances within which comparisons take place shows where power to make the serious world is to be found: who uses it to make a political point, and which ethical dilemmas are implied by the comparison.

Empirical investigations need to discern the parameters of a comparison and situate its social function within the knowledge environment of an utterance. Such 'localisation' in social space, in the intellectual landscape of discourse participants, allows reaching a distance from the comparison itself. Unless such distance is established, as Reno contends in this issue, the dilemma arises that one chunk of knowledge making is claimed to be superior to other chunks of knowledge. According to him, categories are 'social nomads', needing to be situated and understood in their temporal logic when looking for their productive qualities as meaning-makers. It is, thus, through the social functions of crisis knowledge that we can break open the meaning-making at its base.

If the meaning of a comparison can be reconstructed within its empirical setting, the political content of its message can be derived; the political and social purpose of a message can thus be laid open. The positionality of the persons involved is important in this regard: an interventionist class is shaping the discourse, but itself is shaped in its identity by the environment of work, the organisations and political settings within which they bring comparison to the discourse. For example, it makes a difference if the Switzerland comparison is made by a Swiss diplomat of the 1930s or a leading figure of a UN mission tasked with election observation; for the former was an expression of respect while for the latter it works to be apologetic for political aims not achieved. It is not the same if a member of the Taliban or other resistance groups uses a comparison between Afghanistan or Vietnam, or someone in the US forces; for the former, it may be taking inspiration to resist great power repression of national interests, for the latter it may be a warning against trauma and imperial overstretch. In more detail, what could be reconstructed is the differences in meaning-making if someone from the CIA, Marines, or Army uses such a comparison to illustrate their respective institutional perspectives (see Finlan 2014, 60). The question of ethical implications of comparisons becomes accessible by situating and reconstructing the purpose of comparison.

Comparison of conflict in essence needs to be scrutinised through a sociology of the comparer(s). Only by learning something about actors and circumstances of comparisons as much as about the parameters of their becoming applied reveals the role comparison plays in the creation of the 'serious world'. If comparisons are made purposefully, then whose interests, world views, and essentially which problem-perspective do they serve? Unmasking the lack of understanding of complex social dynamics is not a self-serving act of wisenheimery or myth-busting (see Kühn 2016b); de Beauvoir's tension between making sense and making meaning, and Mannheim's view on knowledge as socially situated allows approaching how interventions are pervaded by power and hierarchies. Meaning is ascribed in this power-infested process of claiming exclusive knowledge, suppressing alternative voices. This process reveals the teleological underpinnings within which conflict is understood in general, and how *a* conflict is scaled in particular. This explains how conflict 'solutions' are envisioned and the conflict understood to be the result of a lack that needs be rectified. Hierarchic superiority benefits the interventionist class which profits from the irresolvability of the conflict. It categorises the conflict by comparing it to its knowledge base, creating the meaning required to legitimise its politics.

A note of caution needs to be made for academia, which in no small part is implicated in foreign policy knowledge production: Devising conceptual and methodological approaches which do not stop at designing data bases and mathematical models to transfer the real world into the computers of social science fiction seems advised. Reflectivist and ethically responsible approaches are only just beginning to be developed but continue to be disregarded for their despised complexity and lengthiness by policy makers – what cannot be put on one sheet hardly finds an audience in policy circles. Defying limited attention spans, academic enquiry ought to be inspired by what we find odd, rather than straightforward. That which cannot be compared is what should be put front and centre of our curiosity.

Notes

1. The term 'class' is used her in its widest sense here. As Goetze writes with regard to peacebuilders, a group with large overlap with the intervention professionals referred to here: "The default space of peacebuilding is filled with a specific 'class' of people. … [P]eacebuilders are more easily identified by social characteristics such as economic or cultural capital than by their profession. The notion of class must be understood in a large sense as a wide category of people who exhibit similar sociological characteristics. The boundaries of classes are fluid and fuzzy, and similarity must not be confused with sameness. The trees in a pine forest provide a useful analogy: they may differ in shape, size, and appearance, but they are all still pine trees. In the same fashion, the space of peacebuilding is populated by individuals who in their individual outlook are all different, but who, when considered as a group, share an important number of commonalities, most notably with respect to social origins, education, and their related value structure.' (Goetze 2017, 68).
2. Ruttig's No. 1 is Donald Trump's election campaign claim that inner cities in the U.S. are likely more dangerous than Afghanistan (Trudo 2016) in order to support his electoral campaign priority of ending engagement in Afghanistan.
3. To further illustrate the problematique of analogies, Barfield points out a mismatching comparison involving Swiss cheese: The Swiss cheese model, according to which Afghanistan used to be governed, he argues, describes each slice as unique in that the dispersion of holes is unpredictable; American cheese, on the other hand, is rather homogenous, and implies equal control of the state – its territory and society – in uniform ways. Applying the American cheese model, he argues, would be misleading for Afghanistan: "in other words, when dealing with unevenly distributed resources and populations, it is better to judge the value of your cheese by weight rather than volume, since you know that each slice will have lots of holes (Barfield 2012, 68). Barfield's own point about cheese is, of course, based on ignoring that the US would likely not fit the American cheese model either, given its mixed patterns of inclusion/exclusion, of resources available, uneven control of violence, racism, social stratification etc.
4. With regard to Afghanistan, to be fair, the intervention continues along the lines of building capacity for the fulfilment of a state's core (that is: security) tasks; however, lacking the political structure underpinning administrative

activity, this is a repeatedly frustrated attempt. Yet, the particular Western willingness to get engaged in conflict or post-conflict settings is markedly limited.

5. for example, Melinda Liu 'Ich Bin Ein Hong Konger. How Hong Kong is turning into the West Berlin of the quasi-cold war between the West and China. Foreign Policy, 16 July 2019 available at https://foreignpolicy.com/2019/07/16/ich-bin-ein-hong-konger-protests-china-west-berlin/. Accessed 30 July 2019.
6. To be sure, Tilly himself cautions against using his model for non-Western states (1985, 170).
7. He uses the term not in the commonly shared use of the word establishing 'false belief'.

Acknowledgments

As always, I am deeply indebted to Michael Pugh for helpful comments and insights; Adam Sandor as well as Michael Daxner, Will Reno and Sebastian Büttner as well as two anonymous reviewers all helped sharpen the argument. Thanks also to Bruno Charbonneau and Jonathan Fisher for editorial guidance. The participants of the Comparison Workshop in Montréal (October 2017) and the Käte Hamburger Research Colloquium (July 2019) have significantly contributed by challenging the questions this research has raised. Errors and intellectual shortcomings are my own.

Disclosure statement

No potential conflict of interest was reported by the author.

Funding

This work was supported by the Käte Hamburger Kolleg [Senior Research Fellowship 2019–20].

References

Andersson, R. and Weigand, F., 2015. Intervention at risk: The vicious cycle of distance and danger in Mali and Afghanistan. *Journal of Intervention and Statebuilding*, 9 (4), 519–541. doi:10.1080/17502977.2015.1054655

Barfield, T., 2012. *Afghanistan. A cultural and political history*. Princeton/Oxford: Princeton UP.

Bauman, Z., 1992. *Moderne und Ambivalenz. [Modernity and ambivalence]*. Hamburg: Hamburger Edition.

Bergoffen, D., 2015. Simone de Beauvoir. *In*: E.N. Zalta, ed. *The stanford encyclopedia of philosophy (Fall 2015 edition)*. Available from: https://plato.stanford.edu/archives/fall2015/entries/beauvoir/

BICC GMI, 2017. *Global military index 2017* [online]. Bonn: Bonn International Centre for Conversion. Available from: https://gmi.bicc.de/index.php?page=ranking-table [Accessed 23 August 2019].

Bliesemann de Guevara, B., 2017. Intervention Theatre: performance, authenticity and expert knowledge in politicians' travel to post-/conflict spaces. *Journal of Intervention and Statebuilding*, 11 (1), 58–80. doi:10.1080/17502977.2016.1260208

Bliesemann de Guevara, B. and Kostić, R., 2017. Knowledge production in/about conflict and intervention: finding "facts", telling "truth". *Journal of Intervention and Statebuilding*, 11 (1), 1–20. doi:10.1080/17502977.2017.1287635

Bliesemann de Guevara, B. and Kühn, F.P., 2011. 'The international community needs to act': Loose use and empty signalling of a hackneyed concept. *International Peacekeeping*, 18 (2), 135–151. doi:10.1080/13533312.2011.546082

Bliesemann de Guevara, B. and Kühn, F.P., 2015. On Afghan footbaths and sacred cows in Kosovo: urban legends of intervention. *Peacebuilding*, 3 (1), 17–35. doi:10.1080/21647259.2014.969508

Blumenberg, H., 2006. *Arbeit am Mythos*. Frankfurt am Main: Suhrkamp.

Boltanski, L., 2011. *On critique. A sociology of emancipation*. Cambridge: Polity.

Borger, J., 13 September 2011. The kabul attack shows the insurgency is as potent as ever, the guardian online. Available from: https://www.theguardian.com/world/julian-borger-global-security-blog/2011/sep/13/kabul-taliban-afghanistan [Accessed 19 November 2018].

Bueger, C. and Gadinger, F., 2018. *International practice theory*. Second ed. Cham: Palgrave McMillan.

Calvert, P. and Calvert, S., 2014. *Politics and society in the developing world*. 3rd ed. London/New York: Routledge.

Cassirer, E., 1946. *The myth of the state*. New Haven: Yale University Press.

Cassirer, E., 1979. *Symbol, myth, and culture. Essays and lectures of Ernst Cassirer, 1935–1945*. In: D.P. Verene, ed. New Haven and London: Yale University Press.

Chandler, D., 2010. *International statebuilding: The rise of post-liberal governance*. London: Routledge.

Chandrasekaran, R., 2012. *Little America. The war within the war for Afghanistan*. New York: Vintage.

Cox, R.W., 1981. Social forces, states and world orders: Beyond international relations theory. *Millennium*, 10 (2), 126–155. doi:10.1177/03058298810100020501

Cramer, C., 2006. *Civil war is not a stupid thing. Accounting for violence in developing countries*. London: Hurst.

Daxner, M., 2012. Heimatdiskurs – ein deutsches Problem? *In*: M. Daxner and H. Neumann, eds. *Heimatdiskurs. Wie die Auslandseinsätze der Bundeswehr Deutschland verändern*. Bielefeld: Transcript, 15–68.

De Beauvoir, S., 1948. *The Ethics of ambiguity*. Secaucus: Citadel Press.

De Lombaerde, P., *et al.*, 2010. The problem of comparison in comparative regionalism. *Review of International Studies*, 36, 731–753. doi:10.1017/S0260210510000707

Deutsche Welle, 14 December 2016. 'EU ambassador to Afghanistan "eager to work" with Trump'; interview with Sandra Petermann. Available from: https://www.dw.

com/en/eu-ambassador-to-afghanistan-eager-to-work-wisth-trump/a-36764753 [Accessed 11 July 2019].

Dorronsoro, G., 2005. *Revolution unending. Afghanistan: 1979 to the present*. London: Hurst & Co.

Elsenhans, H., 1981. *Abhängiger Kapitalismus oder bürokratische Entwicklungsgesellschaft. Versuch über den Staat in der Dritten Welt*. Frankfurt/New York: Campus.

Elwert, G., Feuchtwang, S., and Neubert, D., eds, 1999. *Dynamics of violence. processes of escalation and de-escalation in violent group conflicts* Berlin: Duncker and Humblot.

Finlan, A., 2014. *Contemporary military strategy and the global war on terror*. New York: Bloomsbury. et al.

Foucault, M., 2004. *Sicherheit, Territorium, Bevölkerung. Geschichte der Gouvernementalität*. Frankfurt a.M.: Suhrkamp.

Freistein, K. and Gadinger, F., forthcoming. Populist stories of honest men and proud mothers: A visual narrative analysis. *Review of International Studies*.

Gadinger, F., Ochoa, C.S., and Yildiz, T., 2019. Resistance or thuggery? Political narratives of urban riots. *Narrative Culture*, 6 (1) 88–111. doi:10.13110/narrcult.6.1.0088

Gentilini, F., 2013. *Afghan lessons. culture, diplomacy, and counterinsurgency*. Rome and Washington: SSPA/Brookings Institution Press.

Giustozzi, A., 2008. *Koran, kalashnikov and laptop. The Neo-Taliban insurgency in Afghanistan*. New York, NY: Columbia University Press.

Goetze, C., 2017. *The distinction of peace. A social analysis of peacebuilding*. Ann Arbor: University of Michigan Press.

Gramsci, A., 1991. *Gefängnishefte. Kritische Gesamtausgabe in 10 Bänden. Hrsg. v. Klaus Bochmann*. Hamburg: Argument.

Grindle, M.S., 2004. Good enough governance: poverty reduction and reform in developing countries. *Governance*, 17 (4), 525–548. doi:10.1111/j.0952-1895.2004.00256.x

Hameiri, S., Jones, L., and Sandor, A., 2018. Security governance and the politics of state transformation: Moving from description to explanation. *Journal of Global Security Studies*, 3 (4), 463–482. doi:10.1093/jogss/ogy024

Hameiri, S. and Scarpello, F., 2018. International development aid and the politics of scale. *Review of International Political Economy*, 25 (2), 145–168. doi:10.1080/09692290.2018.1431560

Heathershaw, J., 2016. Who are the 'international community'? Development professionals and liminal subjectivity. *Journal of Intervention and Statebuilding*, 10 (1), 77–96. doi:10.1080/17502977.2015.1137395

Hirblinger, A. and Landau, D.M., 2018. Governing conflict: The politics of scaling difference. *Journal of Intervention and Statebuilding*, 12 (3), 385–404. doi:10.1080/17502977.2018.1496644

Hönke, J., 2013. *Transnational companies and security governance. Hybrid practices in a postcolonial world*. Abingdon: Routledge.

Johnson, R., 2011. *The Afghan way of war. culture and pragmatism – A critical history*. London: Hurst.

Jung, D., 2001. The political sociology of world society. *European Journal for International Relations*, 7 (4), 443–474. doi:10.1177/1354066101007004003

Körppen, D., Ropers, N., and Giessmann, H.J., eds, 2011. *The non-linearity of peace processes: Theory and practice of systemic conflict transformation*. Opladen and Farmington Hills: Barbara Budrich Publishers.

Krasner, S., 1994. Compromising Westphalia. *International Security*, 20 (3), 115–151. doi:10.2307/2539141

Kühn, F.P., 2008. Aid, opium, and the state of rents in Afghanistan: competition, cooperation, or cohabitation? *Journal of Intervention and Statebuilding*, 2 (3), 309–327. doi:10.1080/17502970802436338

Kühn, F.P., 2010. *Sicherheit und Entwicklung in der Weltgesellschaft. Liberales Paradigma und Statebuilding in Afghanistan*. Wiesbaden: VS.

Kühn, F.P., 2011. Less is more: International intervention and the limits of external stabilization. *Canadian Foreign Policy Journal*, 17 (1), 62–74. doi:10.1080/11926422.2011.559107

Kühn, F.P., 2016a. The ambiguity of things: Souvenirs from Afghanistan. *Journal of Intervention and Statebuilding*, 10 (1), 97–115. doi:10.1080/17502977.2015.1137397

Kühn, F.P., 2016b. Blumenberg, under-complex analogy and basic myths in international politics: Afghanistan and the 'Graveyard of Empires'. *In*: B. Bliesemann de Guevara, ed. *Myth in international relations: An interpretative approach to studying international relations*. Basingstoke: Palgrave MacMillan, 147–172.

Kühn, F.P., 2017. *Risikopolitik – eine Einführung*. Wiesbaden: Springer.

Lemay-Hébert, N., 2009. Statebuilding without nation-building? Legitimacy, state failure and the limits of the institutionalist approach. *Journal of Intervention and Statebuilding*, 3 (1), 21–45. doi:10.1080/17502970802608159

Lijphart, A., 1999. *Patterns of democracy. government forms and performances in thirty-six countries*. New Haven and London: Yale UP.

Maley, W., 2009. *The Afghanistan wars*. 2nd ed. Houndsmills and New York: Palgrave MacMillan.

Maley, W., 2013. Afghanistan as a cultural crossroads: Lessons from the writings of Ella Maillart, Annemarie Schwarzenbach and Nancy Hatch Dupree. *Asian Affairs*, 44 (2), 215–230. doi:10.1080/03068374.2013.790609

Mannheim, K., 1980. *Strukturen des Denkens*. Frankfurt: Suhrkamp.

Mannheim, K., 2015 [1929]. *Ideologie und Utopie*. Frankfurt: Klostermann.

McAdam, D., Tarrow, S., and Tilly, C., 2001. *Dynamics of contention*. Cambridge: Cambridge UP.

Migdal, J.S., 1988. *Strong societies and weak states. State-Society relations and state capabilities in the third world*. Princeton: Princeton UP.

Migdal, J.S., 1994. 'The state in society. An approach to struggles for domination". *In*: J.S. Migdal, A. Kohli, and V. Shue, eds. *State power and social forces. Domination and transformation in the third world*. Cambridge/New York: Cambridge UP, 7–34.

Paris, R., 2004. *At war's end. Building peace after civil conflict*. Cambridge: Cambridge UP.

Pouligny, B, 2006. *Peace operations seen from below: UN missions and local people*. London: Hurst.

Rubin, B.R., 2002. *The fragmentation of Afghanistan. State formation and collapse in the international system*. New Haven/London: Yale UP.

Ruttig, T., 18 September 2010. All together now: 'This is not Switzerland'. Available from: https://www.afghanistan-analysts.org/2010-elections-6-all-together-now-this-is-not-switzerland/ [Accessed 19 November 2018].

Ruttig, T., 2017. Brasilien … oder: „Hört auf mit dem Rumgeheule": Die Top 5 der verharmlosendsten Afghanistan-Vergleiche; online https://thruttig.wordpress.com/2017/07/30/brasilien-oder-hort-auf-mit-dem-rumgeheule-die-top-5-der-verharmlosendsten-afghanistan-vergleiche/ [Accessed 19 November 2018].

Saikal, A., 2004. *Modern Afghanistan. A history of struggle and survival*. London/ New York: I.B. Tauris.

Schneckener, U., 2005. *Die soziale Konstruktion des ‚Terrorexperten'. Terrorismusforschung zwischen Medienlogik und Politikberatung". Diskussionspapier der FG 8, 2005/03 (Mai 2005)*. Berlin: SWP.

Sending, Ole Jacob, 2015. *The politics of expertise. Competing for authority in global governance*. Ann Arbor: University of Michigan Press.

Shahzad, S.S., 2011. *Inside Al-Qaeda and the Taliban. Beyond Bin Laden and 9/11*. London: Pluto.

Small Arms Survey, 2018. *Civilian firearms holdings 2017* [online]. Available from: http://www.smallarmssurvey.org/fileadmin/docs/Weapons_and_Markets/Tools/Firearms_holdings/SAS-BP-Civilian-held-firearms-annexe.pdf [Accessed 27 December 2018].

Smirl, L., 2015. *Spaces of aid. How cars, compounds and hotels shape humanitarianism*. London: Zed.

Suhrke, A., 2011. *When more is less. The international project in Afghanistan*. New York: Columbia University Press.

Tilly, C., 1985. War-making and state-making as organised crime. *In*: P.B. Evans, D. Rueschemeyer, and T. Skocpol, eds. *Bringing the state back in*. Cambridge: Cambridge UP, 167–191.

Trudo, H., 2016. Trump calls Afghanistan safer than America's inner cities; politico 09/12/2016, Available from: http://www.politico.com/story/2016/09/trump-afghanistan-inner-cities-228057 [Accessed 19 November 2018].

von Trotha, T., 1997. 'Zur Soziologie der Gewalt.' [On the sociology of violence]. *In*: T. von Trotha, ed. *Soziologie der Gewalt. Sonderheft Kölner Zeitschrift für Soziologie und Sozialpsychologie*. Opladen: Westdeutscher Verlag, 9–56.

Woodward, S.L., 2017. *The Ideology of failed states. Why intervention fails*. Cambridge/ New York: Cambridge UP.

Sahelistan? Military Intervention and Patronage Politics in Afghanistan and Mali

Romain Malejacq and Adam Sandor

ABSTRACT

The 'Sahelistan' discourse that conflates conflict dynamics in Afghanistan and Mali is widespread in Western media and policy circles. We argue that such representations contribute to the adoption of one-size-fits-all conflict management policies. Drawing on extensive fieldwork in both places, we also argue that these standard templates of intervention shape responses from local non-state armed actors, who manipulate foreign interveners for their own (violent) purposes in similar ways. Yet, we show that this convergence in armed group behaviour still manifests itself in important variations in widely different contexts, furthering strongmen autonomy in Afghanistan, while fostering armed group fragmentation in Mali.

Western media and policy makers frequently fuse understandings of conflict dynamics in areas of limited statehood, where armed insurgencies and identity-based politics often prompt external military intervention. A recurrent, prominent case of conflict amalgamation is the comparison made between the war in Mali since the beginning of French military operations in 2013 and the one in Afghanistan since the start of the U.S.-led intervention in 2001. International security experts and journalists now commonly consider the Malian conflict as a 'new' or 'second' Afghanistan for Western militaries (Cecchinel 2017, Lema Landu 2018). 'Mali is our Afghanistan,' wrote a *Le Monde* reporter in 2017 (Ayad 2017).[1] Afghanistan, in turn, is viewed as a case to learn from for those foreign armies willing to intervene in the Sahel (Philp 2020).

The comparison is not only made by pundits and journalists in mainstream media and flashy news outlets. During the first decade of the 2000s, U.S. counter-terrorism activities in the Sahel and the Sahara were already underpinned by the idea that the broader region would become 'the next Afghanistan' (Schmidle 2009). In 2012, referring to Al-Qaeda in the Islamic Maghreb (AQIM)'s presence in northern Mali, French Minister of Foreign Affairs Laurent Fabius proclaimed the advent of what he called 'Sahelistan' (2012a): a combination of 'the Sahel, a strip of arid territory bigger than France, and the terrorism that affected Afghanistan' (2012b). The following year, U.S. Secretary of State Hillary Clinton likened the possibility of AQIM developing a safe haven in Mali to Al-Qaeda's use of Afghan territory in the planning of the 9/11 attacks (see McGreal 2013) while British Prime Minister David Cameron urged the world 'to come together to deal with this threat in North Africa,' just as it had in Afghanistan and Pakistan (quoted in Lewis 2013). Western military officials and top diplomats have made similar comparisons. A U.S. Africa Command strategist stated that the United States' military posture in the Sahel aimed to 'prevent "Afghanistan-like effects"' (quoted in McNeill 2017, p. 50). French Colonel Michel Goya argued that avoiding the mistakes made in Afghanistan formed a guiding principle of France's intervention in Mali (see Erforth 2020, pp. 79–80). And current United States Special Envoy for the Sahel, J. Peter Pham, has pointed out the 'uncanny parallels' between Afghanistan and Mali (2013).

Conflicts where local Islamist groups become intertwined with transnational Salafi jihadist organisations have become representative and symptomatic of the so-called Global War on Terror. Seen from Paris, London, and Washington, places where weak state capacities blend with ethnic, tribal, or clanic divisions, insurgent violence, and Islamic militantism are often reduced to different manifestations of the same global phenomenon. Yet, in Mali and Afghanistan, only the broadest conflict features are consonant. As a former French Defence official for Mali told one of us: 'Officials back in Paris often say to me, "so what you're saying is the conflict in Mali is like Somalia or Afghanistan, right?" And I reply, "No! No, it's nothing like that!" I'm always telling people this. It is completely different.'[2]

The conflation of armed conflicts into broader, more encompassing categories is nothing new. A well-established body of literature in Security Studies and International Relations has shown, for example, how representations about so-called failed states drive our very understanding of postcolonial statehood and guide foreign policy (Bilgin and Morton 2002, Jones 2008, Woodward 2017). What we are witnessing today with analyses conflating conflict dynamics in Mali and Afghanistan is only the latest variant of a tendency to over-simplify the apparent messiness of political violence in the Global South. Scholars have advanced a number of explanations for this, from the news media's structural inability to deal with complexity (Minear *et al.* 1996) and the pervasive influence of simplistic and all-encompassing theories of conflict (Malejacq 2010), to the need for and reproduction of 'simple narratives' (Autesserre 2012). In this article, we do not attempt to provide a causal explanation of why conflicts that only share broad features are so often boxed together. Neither do we intend to condemn all comparative endeavours. Careful, controlled, academic comparison is useful and important.[3] Rather, we aim here to stress that fusing conflict knowledge regarding broadly similar situations of asymmetrical warfare, rebellion, and counter-terrorism has political effects, notably with regard to how international interventions are conceptualised and how they unfold in interactions with local actors.

We start from the premise that discourses and representations matter. More specifically, they matter for foreign policies, which depend 'upon representations of the threat, country, security problem, or crisis they seek to address' (Hansen 2006, p. 6). It is through those representations that policy makers develop prescriptions for how to deal with a given problem. War analogies like 'Sahelistan' matter because they 'perform diagnostic or inferential tasks,' thereby framing how decision makers evaluate between multiple alternative options (Khong 1992, pp. 252–253). In fact, such careless conflation of conflict patterns strips context from analysis, which necessarily hinders comprehension.[4] On this account, analogical comparisons, even when done casually, perform consequential political labour. They produce 'categories of perception' that simplify and naturalise these diagnoses, and make thinking differently a heretical venture (Bourdieu 1991, pp. 127–132).

Certainly, representations of threat matter all the more when they are evoked by heads of state, government ministers, and military strategists. But media outlets and so-called experts also contribute to the legitimisation of these discursive formations. So does academic popularisation. After all, 'leaders do not act in a vacuum: rather their utterances are made possible by existing societal discourses' (Fierke 2007, p. 104). Indeed, the 'Sahelistan' discourse is pervasive and widespread.[5] When such categories of conflict knowledge become hegemonic, thinking differently about how to respond to the complexity of armed conflicts becomes a challenge for policy makers, thereby shaping 'conflict management policies themselves' (Charbonneau and Sandor 2019, p. 439). As a result, such comparisons contribute to the adoption and implementation of one-size-fits-all solutions to what decision makers analyse as manifestations of a same phenomenon. Hence, interventions grounded in these comparisons often reduce the resolution of complex, idiosyncratic, political dynamics to disastrous, technical problem-solving approaches. Representations that leverage comparisons of armed conflicts reinforce categories of threat, and are thus far from neutral.

We claim that the '"Sahelistan" terrorism narrative' (Pérouse de Montclos 2015) reflects the nature of local responses to international military intervention more than the nature of the armed conflicts themselves. We argue that it has been the ability of local non-state armed actors to leverage foreign interveners' practices and resources which has been the enduring (and comparable) pattern for conflict parties seeking to maximise power in Afghanistan and Mali. In particular, it is the similar ways in which external actors have advanced conflict resolution through military action and the massive infusion of resources that has shaped responses from domestic actors in similar ways. This tells us more about how local groups and individuals come to subvert global opportunity structures, practices, and forms of intervention than it does about how they understand and pursue political objectives through violent action. Finally, we also show that, despite this increasing convergence of practices, the groups and individuals operating in Afghanistan and Mali still adopt distinctive modes of behaviour that reflect how fundamentally different these two conflicts were in the first place.

In this article, we build on cumulative knowledge of political violence in Afghanistan and Mali and qualitative data collected over years of fieldwork conducted separately (and for unrelated projects) in these two countries. One of us has spent over 12 months in Afghanistan over the last 13 years. Since 2007, he has interviewed over two hundred high-profile political actors (many multiple times). The actors he has interviewed include warlords and their entourages (advisers and family members), current or former presidents, vice presidents, ministers, governors, intelligence officers, opposition leaders, mujahideen, foreign representatives, NATO and UN officials, NGO workers, journalists, local researchers, and more. Since 2012, the other has spent 16 months interviewing conflict actors in Sahelian capitals (Bamako, Ouagadougou, and Niamey) as well as in areas controlled by armed groups (mostly in the Timbuktu region). He has conducted interviews with armed group leaders, rank-and-file fighters and their families, jihadist group defectors, government ministers, drug traffickers and smugglers, religious elites, journalists, diplomats, UN staff, and numerous others. While a great number of our interviews were semi-structured, we have engaged in many more lengthy, informal, repeated conversations over meals or drinking tea. Addressing the trust deficit inherent in conflict areas has indeed involved building networks and relationships, spending significant bouts of time with interlocutors, observing formal and informal political gatherings, and engaging in long-standing digital conversations over the years (see Malejacq and Mukhopadhyay 2016). The data outlined below, while carefully triangulated by using primary material and by confirming information via multiple, unrelated sources, reflects our own positionality. It is necessarily limited by conditions of geographical and social access (which are highly gendered in Afghanistan and Mali), our citizenships, and questions regarding interlocutor, broker, and researcher safety.[6]

The remainder of the article proceeds in four parts. We first provide a brief overview of the conflicts in Afghanistan and Mali, emphasising key historical, sociological, and political differences between the two. Next, we illustrate some of the most important ways in which Western interveners have adopted and implemented similar policies in both places. In the third section, we identify and compare three ways in which local actors instrumentalise external interveners' policies to suit their interests and take advantage of newly-created global opportunity structures. In the conclusion, we circle back to how differences in contexts lead to variation in processes and outcomes and highlight how similar policies further strongmen autonomy in one case, Afghanistan, and foster armed group fragmentation in the other, Mali.

A Quick Glance at Conflict in Afghanistan and Mali

The conflicts taking place in Afghanistan and Mali share similarities: they are both cases of jihadist insurgencies operating in places where 'a spectrum of conventional bureaucratic state capabilities [exist] alongside (generally very strong) informal political networks' (Reno 1998, p. 2) and within a multi-ethnic, fragmented society. Yet, Afghanistan and Mali are fundamentally different, historically, geographically, socio-culturally, or even in terms of geopolitical position in their respective regions, and so are the conflicts taking place in their midst. Though the two conflicts might, at first glance, look alike, they are actually very distinct. Here we do not aim to delve into the fundamental differences that exist between the two societies. Rather, we identify at least four important ways in which the conflicts themselves differ that make any out-of-context comparison potentially risky.[7]

First, the conflicts in Afghanistan and Mali are incomparable in terms of duration and scale. Initially the product of severe Cold War rivalry, the Afghan war has continued, in its different iterations, for over forty years, including a decade-long Soviet occupation in the 1980s and a gruesome civil war in the 1990s. Four decades of war have produced horrific conflict legacies: over a million people killed, millions of refugees and internally displaced persons, ravaged infrastructures, a decimated economy, and an entrenched history of radicalisation. In comparison, Mali has experienced only four, relatively short periods of rebellion since the country's independence in 1960: in 1963–1964, between 1990 and 1996, briefly in 2006 (followed by a lingering, low-level insurgency until 2009), and since 2012. Consequently, casualties and the destruction of national infrastructures resulting from these armed rebellions are nowhere near as high as in Afghanistan. In addition, unlike in Afghanistan, armed rebellion and political violence in Mali have been confined geographically. The 1963–1964 and 2006 periods of rebellion were mostly limited to one of Mali's most northern regions (Kidal) while the fighting that occurred between 1990 and 1996 hardly extended beyond Mali's three northern regions (Gao, Kidal, and Timbuktu). Only during the current crisis has armed conflict migrated towards the centre of the country. Thus, since Mali's independence, armed violence has remained concentrated in sparsely populated central and northern areas, largely sparing the majority of the country's economic infrastructures located in the South. In turn, the scale of foreign military involvement in Mali (and the Sahel more broadly) is in no way comparable to the post 9/11 U.S.-led intervention in Afghanistan. At its peak, in 2011, foreign military presence in Afghanistan reached up to 140,000 troops. Foreign military forces in Mali only represent a small fraction of that – around 12,000 peacekeepers as part of the United Nations Multidimensional Integrated Stabilisation Mission in Mali (MINUSMA) and slightly over 5,000 French troops as part of *Opération Barkhane*.[8]

Second, limiting our understanding of the armed conflicts in Afghanistan and Mali to cases of jihadist insurgencies operating in and against weak and corrupt states is grossly misleading. Ranked 9th and 33rd respectively in the most recent Fragile States Index (2020), the Afghan and Malian states do not have much in common empirically. Afghanistan has arguably suffered near complete state collapse in 1992, when the vestiges of state institutions that remained from the war with the Soviet Union (like the armed forces) were dismembered during fierce competition over state capture (see Rubin 2002, pp. 265–280). Built on a decimated state apparatus, today's Afghan state stems from an ambitious but critically flawed, 'project of social engineering' (Suhrke 2007, p. 1291) and has been largely incapable of providing for its population in the post-Taliban era. Power in Afghanistan is greatly personalised, warlords remain highly influential (Malejacq 2019), and political elites have reaped the benefits of the U.S.-led intervention for close to two decades. In contrast, at no point in post-independence Malian history have the country's institutions been as gutted as those of Afghanistan. At no time has the Malian Armed Forces completely dissolved. This is not to say that Mali's state institutions are efficient in the provision of public services, or that even territories free of armed rebellion have not been undergoing a steady process of 'empirical state failure' (Bleck and Michelitch 2015). Some even argue that 'in many ways, Mali's empirical existence as a state has been deficient since its inception in 1960, and attempts at constructing it have always been unsuccessful' (Craven-Matthew and Englebert 2018, p. 8). Nevertheless, there is a difference between state institutions marked by personalised and corrupt practices, incompetent capacities, and structural deficiencies, like those in Mali, and those that have been essentially destroyed by over forty years of ongoing warfare over the vast majority of its national territory, such as those in Afghanistan. The reach and capacity of Malian state institutions are marked by drastic regional, and urban/rural variance, but they have never collapsed entirely.[9]

Third, the enemies these states and their external backers face are fundamentally different. The main Islamist groups operating in Mali, fighting under the banner of the *Jama'at Nasr al-Islam wal Muslimin* (JNIM) and the Islamic State in the Greater Sahara (ISGS), bear little resemblance with the Taliban, the organisation that unmistakably dominates the Afghan insurgency – despite the presence of other groups such as Al-Qaeda and the Islamic State - Khorasan Province (IS-KP). Conflating the two conflicts on the basis that insurgents of both countries claim to be fighting a jihad against an illegitimate, heathen government is overly simplistic. First, these groups have completely different origins and political trajectories. The Taliban is a politico-military organisation that was created in the mid-1990s, conquered and controlled most of the Afghan territory with the help of Pakistani military intelligence in the second half of the 1990s, and was toppled by the U.S.-led intervention in 2001. It then re-emerged in the form of an insurgency in the mid-2000s – still with the support of Pakistan, whose territory Taliban fighters have used as a safe haven ever since. The organisation promotes a nationalistic agenda that aims to establish an Islamic state following sharia law in Afghanistan and Afghanistan only.[10] It is predominantly Pashtun in composition, and hence represents the country's strongest, historically dominant, ethnic group, which confers on to the organisation a degree of popular support in the areas in which it operates (as well as fosters a degree of resistance among other ethnic groups). In contrast, JNIM consists of a fluid coalition of armed Islamist groups, composed of several Sahelian ethnicities, that operate in relatively autonomous brigades.[11] While enjoying some local legitimacy, JNIM's precursor organisations had nowhere near such influence in or support from communities in northern or central Mali, let alone in the rest of the country, as have the Taliban in Afghanistan. Moreover, while JNIM's military operations and governance practices, like the Taliban's, are fundamentally local and centred on survival and the maintenance of sharia law in spaces under their immediate control, their stated agenda is decidedly transnational: the establishment of an Islamic caliphate across the Sahel and the Sahara (Bencherif 2020).

Finally, the war economies of Afghanistan and Mali drastically differ. While media outlets and some academics and policy-makers place the role of drug trafficking front and centre in their framing of both conflicts, a cursory examination of the impact of the drug economy on conflict dynamics in the two countries demonstrates the dissimilarities between the cases. First, Afghanistan is a drug-producing country, which, in the past five years, accounted for 84% of the world's illicit cultivation of opium poppy (United Nations Office on Drugs and Crimes 2020, p. 9), whereas Mali boasts a comparatively very modest production of cannabis in the south of the country. The Afghan drug economy permeates most layers of state and society and, consequently, deeply affects the conflict dynamics and the strategies of all the actors involved in it. On the contrary, Mali's cannabis production shows no evidence of impacting the conflict occurring in the country's north. While its northern Saharan territory is indeed an important space of transit for Moroccan hashish (and, to a lesser extent, for Latin American cocaine destined for Europe),[12] the number of people that actively take part in drug trafficking in northern Mali pales in comparison to the sheer geographic scale of economic opportunity and associated livelihood structures that opium production and trafficking provide in Afghanistan.[13] Second, the Taliban actively sponsor and profit from drug production (through taxation and protection) and drug trafficking, with some commanders even directly engaging in heroin processing (Peters 2009). This is in stark contrast to the activities of armed Islamist groups operating in the Sahel, who have rarely been directly involved in organising trans-Saharan narcotics smuggling networks. Sahelian security officials, businessmen, and traffickers themselves agree that armed Islamist groups operating in northern Mali are not directly involved in drug trafficking, but only tax traffickers passing through their zones of control.[14] One should certainly avoid uncritically pronouncing the existence of a 'drug-terror nexus' (Pham 2010) and 'narco-terrorist' groups in the Sahel (see Lacher 2013).

Intervention Dynamics

Foreign policy initiatives designed to respond to civil war situations almost inevitably draw on universalised templates familiar to most interveners (see Paris 2004, Mac Ginty 2012, Autesserre 2014). We highlight below three comparable ways in which intervening actors have operated in Afghanistan and Mali.

The first, maybe most obvious, similarity between the external involvements in Afghanistan and Mali is the almost unquestioned resort to military action as the main, at times only, answer to the threat posed by Salafi jihadist armed groups. In both cases, interveners proved to be incapable of solving the conflicts' root causes through military action, despite successful initial involvement. In Afghanistan, the United States reacted to a series of coordinated attacks on American soil planned and executed by a transnational terrorist network (Al-Qaeda) by launching a military intervention openly aimed at changing a regime that the U.S. administration considered a threat to international security (the Taliban). 'We will make no distinction between the terrorists who committed these acts and those who harbour them,' asserted George W. Bush in his address to the nation on 11 September 2001. The regime was swiftly toppled. The remaining Taliban either disappeared by leaving the battlefield or by finding refuge in Pakistan, but an insurgency erupted only a few years later. Two decades later, U.S. soldiers are still present in Afghanistan, and the insurgents remain undefeated. In Mali, in January 2013, France launched *Opération Serval*, a military intervention aimed at halting the advance of a jihadist force that had previously taken control over three of the country's northern regions (Gao, Kidal, and Timbuktu) and was now moving southward towards the capital. The mission was a short-term success: the French quickly retook all major towns in the North, killed an estimated 500 fighters, and stopped the battalions pushing south, driving back those that remained to northern Mali (Wing 2016). However, *Serval* failed to address the widespread political marginalisation and neglect of several northern communities, from which many of these fighters originate (Charbonneau and Sears 2014, p. 208).

In their retreat, jihadist fighters found shelter in neighbouring countries, formally entered existing armed groups or pro-Malian government militias, or rejoined their families and communities in rural areas. Their organisational fluidity and pragmatic orientation allowed these armed Islamist groups to soon thereafter reconstitute themselves as a resilient insurgency (Sandor and Campana 2019).

The second consonant feature of the interventions in Afghanistan and Mali, logically following from the first, is that both, though initially conceived as limited in their scopes and ambitions, were progressively drawn into mission creep. In both cases, the interveners' inability to solve the conflicts' root causes through military action led them to fall into investment traps wherein initially small military involvements turned into highly ambitious and long-lasting missions. In Afghanistan, the U.S.-led military intervention was initially conceived as a short-term operation combining regime change and counterterrorism. '[I]ndigenous allies replace[d] American conventional ground troops by exploiting U.S. airpower and small numbers of American special operations forces (SOF)' (Biddle 2005-2006, p. 161), which then allowed their Afghan allies to capture most state institutions (Malejacq 2015). This military strategy, combined with a 'light footprint' approach to Afghan political affairs, was meant to ensure minimal U.S. and UN involvement (Chesterman 2002). Yet, this initial purpose rapidly evolved into an ambitious and dysfunctional state-building mission combined, after the rise of the Taliban insurgency, with the adoption of a counterinsurgency strategy and an extensive military surge starting in 2009 (Suhrke 2011). In Mali, *Serval*'s military aims were initially quite modest also: 'destroy[ing] ... and disorganiz-[ing] the terrorist network' that was pushing towards Bamako (Assemblée Nationale 2013), and preserving the integrity of the state.[15] Neither French officials nor their international counterparts favoured a large state-building project or a long-lasting military intervention (Charbonneau and Sears 2014, p. 196). However, by the summer of 2014, the French military intervention was transformed into an ambitious counter-terrorism endeavour aimed at fighting jihadist groups in the whole Sahel region – *Barkhane*. Meanwhile, the role of MINUSMA is to support the extension of the Malian state's authority over its entire territory. Given the complexity of the task, MINUSMA officials now assume their presence in Mali will last for the next twenty to thirty years,[16] while French military officials now admit that they were initially 'exaggeratedly over-optimistic' (quoted in Lagneau 2019).

The third similarity is that, in both Afghanistan and Mali, a major priority of the interveners was to establish a post-war governance order following the signing of a power-sharing agreement. As such, these cases are exemplary of how power-sharing agreements have become the footings of the institutional scaffolding meant to usher in post-war reconstruction efforts in multi-ethnic, fragmented, societies. In both cases, military intervention paved the way for armed groups to assume government functions and become legitimised while others, armed Islamist groups in particular, retained their classification as illegitimate, and were excluded from the political process – at least, in the case of Afghanistan, until recent talks between the U.S. administration and members of the Taliban. In Afghanistan, the Taliban were entirely excluded from the Bonn conference of November-December 2001 that would decide on the country's political future. Regime change in the direct aftermath of 9/11 was incompatible with broad-based reconciliation: remnants of the Taliban regime were fought on the battlefield or pushed across the Afghanistan-Pakistan border; Al-Qaeda operatives were killed or captured; while representatives of other Afghan factions – some of which were fighting the Taliban alongside the U.S. military – and political elites received the lion's share in the new political order. The political marginalisation of the Taliban and, de facto, of the Pashtun communities that supported them, in large part explains the rising insurgency that followed. In Mali, armed groups willing to officially distance themselves from Al-Qaeda affiliates were included in the peace negotiations that would establish a power-sharing agreement for the country. Local fighters that had directly participated in jihadist rule in northern Mali thus became full-fledged signatories of the 2015 Accord for Peace and Reconciliation (APR). Those who refused to do so were systematically excluded from formal peace negotiations and have been targeted by *Barkhane* ever since (Bouhlel 2020).

Instrumentalisation and Adaptation

It is the universalised templates identified above, we argue, that alter patronage politics and incentivise local actors to instrumentalise international interveners in similar ways. Elements of the international system indeed push non-state armed actors to adopt seemingly treacherous behaviours as they attempt to resist international pressure to conform. However, this is not to say that there is no variation in the way non-state armed actors behave. On the contrary, we show that, despite externally-induced convergence in their behaviours, Afghan and Malian armed groups remain fundamentally different. We illustrate this by analysing and comparing three areas in which external military intervention has shaped conflict dynamics in Afghanistan and Mali: the local political economy of violence, the instrumentalisation and creation of security knowledge, and 'post-war' governance institutions.

Local Political Economy of Violence

Maybe the most flagrant, yet seemingly indecipherable, way in which external military intervention affects non-state armed groups is by creating new political incentive structures and altering old ones, thereby transforming (and further complexifying) the local political economy of violence (Ahmad 2018).

Armed group leaders operate in unstable environments that foster self-centred behaviours and short-term political and material gains (Reno and Matisek 2018). Newly-introduced global opportunity structures prompt them to adopt duplicitous behaviours in the eyes of the international community, such as splintering[17] and side-switching.[18] These behaviours can be observed in both Afghanistan and Mali and create an apparent chaos that leads pundits and policy makers to box these conflicts (and others) together. However, closer examination of the two conflicts shows that local power brokers' adaptation to external intervention, which occurs mainly through side-switching and repositioning in Afghanistan, plays out quite differently in Mali, where armed group fragmentation along community/clan-based lines has become the norm, far outstripping side-switching.

In Afghanistan, Abdul Rashid Dostum is certainly considered the epitome of opportunistic side-switching. He has been changing colours for the best part of the past four decades: from pro-government militia leader during the Soviet Afghan war of the 1980s, to warlord in control of his own fiefdom in the early 1990s, to Taliban challenger, to American ally in the early days of the U.S.-led intervention, to political pariah in the years that followed, and finally to first vice president of the Islamic Republic of Afghanistan between 2014 and 2020. Dostum has switched between many different patrons, from Russia and Uzbekistan (and before, the Soviet Union) to Iran, Pakistan, Turkey, and the United States; he is often portrayed by Western media and policy makers as a compulsive opportunist prone to betrayal; but, for many Afghan Uzbeks, he represents continuity in an otherwise fluid environment. His apparent fickleness has in fact allowed him to adapt to changes in international environments and opportunity structures, and hence survive. In 2001, he provided the United States with a powerful ally on the ground to fight the Taliban. Incidentally, U.S. support gave him the opportunity to rearm, remobilise, and reaffirm his authority in the north of the country. Rais Baghrani, a commander from southern Afghanistan, is another good example. Originally a member of the hard-line faction of the Afghan communist party, he successively affiliated himself with a number of mujahideen groups during the Soviet-Afghan war, then with the Taliban, and eventually with the Karzai administration after 2001. 'Each change in organisational "membership,"' wrote former British Army officer Mike Martin, 'was due to evolutions of his local political context, which he needed to either exploit or not be destroyed by' (2014, pp. 54–55). Opportunity structures have also fluctuated in the midst of intervention itself. Derksen for example shows how Dostum's forced exile in 2017 led some of his men to join the ranks of the Taliban or IS-KP (2017).[19] 'With Dostum absent,' she notes, 'resources for those commanders he had mobilized against the Taliban in Faryab in 2015 and 2016 again dried up' (*ibid.*). Derksen notably mentions the case of Mohammad Ashraf and his men who, after not receiving their salaries for half a year, eventually decided to switch sides and join the Taliban. 'He now calls himself Mullah Ashraf,' she writes (*ibid.*).

In Mali, the external military intervention and the 'post-war' governance order that followed have spurred armed group fragmentation and a rapid proliferation of new non-state armed groups. An important grievance that has informed northern Malian armed groups stems from minimally-implemented political reforms regarding the control of local administrations established in peace agreements from the 1990s. To address this grievance, the APR involves widespread decentralisation measures, including the creation of new administrative regions, which have fostered an environment of spoil-seeking and uncertainty, and prime political economies for armed violence.[20] Since the onset of peace negotiations in 2013, local administrative positions have become especially coveted by the communities that form the backbone of Mali's armed groups. The creation of the *Mouvement pour le Salut de l'Azawad* (MSA) by Moussa Ag Acharatoumane of the Daoussahak tribe is illustrative. Acharatoumane – though a founder of the separatist *Mouvement National de Libération de l'Azawad* (MNLA) – launched the MSA in September 2016 (Desgrais *et al.* 2018, pp. 670–71). Daoussahak notables and MSA fighters insist that one of the predominant reasons the MSA was created was to secure administrative positions and ensure a higher degree of tribal representation in the newly-created Menaka region.[21] The creation of the MSA, however, stoked tensions amongst the Daoussahak, leading to regular skirmishes between the group and the remaining Daoussahak holdouts that stayed loyal to MNLA allies – the *Haut Conseil de l'Unité de l'Azawad* (HCUA) – or to local Daoussahak jihadist battalions (Dubois 2018). MSA fighters accuse Daoussahak holdouts of maintaining ties with jihadist groups, while the latter accuse Acharatoumane of treachery. The creation of the *Congrès pour la Justice dans l'Azawad* (CJA) provides another example. In 2016, fighters of the *Coordination des Mouvements de l'Azawad* (CMA) belonging to the Kel Ansar tribe grew frustrated by their insufficient inclusion in APR's institutional bodies. Arguing that the 'scraps' the CMA leadership allotted to their community failed to account for their tribe's significance in the Timbuktu region, a majority of the CMA's Kel Ansar members baulked at the group's allocation of positions to local Arab community rivals. Kel Ansar fighters subsequently left in droves to create the CJA.[22] Tensions came to a head the following year when CJA fighters took control of Malian Army checkpoints throughout Timbuktu to stop state officials from entering the city until their demands were met, laying siege for 45 days.

Both the Mali and Afghanistan cases demonstrate that the international community's efforts to devise 'post-war' orders create opportunities to acquire newly-introduced internationalised rents that alter existing patronage politics and produce convergence in armed group behaviour. Yet, we still observe contextually-significant variations. In Afghanistan, responses to globalised incentive structures manifested themselves in local strongmen pursuing opportunistic side-switching to stay relevant and survive as the security context changed. Conversely, when international resources are divvied up in a contested fashion, as in Mali, pressure is placed on armed group coalitions which, in that case, resulted in widespread armed group fragmentation along community lines.

Instrumentalisation and Creation of Security Knowledge

When Western interveners abide by universalised conflict categories and intervention formulas informed by Sahelistan-like discourses, they place themselves at a disadvantage in their interactions with local armed groups. Interveners' knowledge is necessarily ambiguous and incomplete. This allows armed groups to position themselves as local experts and brokers with insider knowledge that interveners need but have a difficult time placing in context. As a result, armed groups leverage their position to offer up information which can be radically simplified, exaggerated, or patently fabricated and framed in such a way to support their political position at the expense of local enemies. In other words, universalised grids of intelligibility are instrumentalised by local armed entrepreneurs who create security knowledge to suit their own (violent) purposes. Once again, these behaviours can be observed in both Afghanistan and Mali. Closer observation instead shows that Afghanistan's armed groups have been instrumentalising foreign interveners for a much longer time, and at a much larger scale, than the ones in Mali, in large part because they hold considerable sway in their communities and have a far greater ability to mobilise followers than their northern Malian counterparts.

In post-2001 Afghanistan, examples of power brokers who have successfully instrumentalised their relations with foreign interveners abound. In the early days of the post 9/11 intervention, Sher Mohammad Akhunzada, a former mujahid turned provincial governor, manipulated U.S. and British ignorance of the southern Afghan terrain to eradicate his rivals' poppy fields. He then leveraged his proximity to the United States (and American money) to solidify his position in the local political economy and extract additional resources from development projects (Martin 2014, pp. 23, 120, 129, 134). In the southern province of Kandahar, Governor Gul Agha Shirzai operated in a similar way, taking advantage of U.S. unfamiliarity with the local context, need for allies, and largesse to cultivate and expand his patronage networks and further establish his authority. In Chayes' words: 'U.S. forces were helplessly wrapped inside the Shirzais' friendly bear hug' (2006, p. 181). Similar patterns of action also took place at more localised scales of politics: in the eastern province of Kunar, for example, U.S. Army Special Forces units initially relied on outdated, Orientalist interpretations of Afghan tribal structures that played directly into the hands of local brokers seeking to buttress their dominance. These 'entrepreneurs who fed the U.S. Army with auxiliaries and intelligence' were hence provided with the opportunity to acquire contracts and secure their positions as distributors of large sums of money in exchange for their efforts to exclude the Taliban from town and village activities (Baczko 2016, pp. 1421–1422). In Afghanistan, the very skilled have even created disorder, fear, and unpredictability to increase the requirement for those who can deliver political stability or get rid of rivals now deemed incapable of doing so. In the southern province of Helmand, Mir Wali, one of the local warlords, was in charge of the highway police. In 2004, he responded to his replacement by a private security company by attacking the checkpoints he used to control, meanwhile claiming that these violent incidents were in fact the Taliban's doing (Martin 2014, p. 141). In 2008, in the same province, local political brokers reportedly organised a fake Taliban attack to 'create chaos and prove they were the only people who could lead the province' (Habibullah Khan, district governor of Nad-i Ali, Helmand Province, quoted in Martin 2014, p. 178).

In Mali, aspiring strongmen similarly instrumentalise French forces by accusing rivals of 'being in league with the terrorists' in order to acquire a rival's property, consolidate their authority, or settle old scores.[23] For example, in Kidal, in the early period of the French intervention, Mohamed Ag Najim, the MNLA's top military official, used his partnership with *Serval* to gain the upper-hand against his rivals in the illicit drug trade. In early 2013, he informed *Serval* officials of an armed convoy of jihadists moving towards the Algerian border near the town of In Khalil. French fighter jets bombed the convoy, discovering later that it was composed of Arab businessmen that were embroiled in a turf war with Najim over control of this border zone (International Crisis Group 2018).[24] To consolidate their political authority and leadership over the Timbuktu region's Tuareg confederations and establish control over refugee camps in Mauritania, notables of the Kel Ansar tribe, Abass Ag Mohamed Ahmed and Abdoul Majid Ag Mohamed Ahmed (also known as 'Nasser'), leveraged their relationships with *Barkhane* officers to sideline challengers both within their tribe and in rival ones, the Kel Ghazzaf, Idakakamane, and Cherifien.[25] As a result, supporters and fighters of the HCUA and other armed groups from these tribes are regularly suspected by the French of supporting local jihadists, and several have been arrested or even killed in *Barkhane* operations. In the Menaka region, leader of the *Groupe d'Autodéfense Touareg Imghad et Alliés* (GATIA), General El Haji Ag Gamou, secured a partnership with Barkhane to use the militia as a counter-terrorist force. To settle personal scores and establish himself as Gao and Menaka's central politico-military figure, Gamou has since provided *Barkhane* intelligence on rival Fulani communities, insisting (not inaccurately) that armed Islamist groups had been recruiting heavily among them. Gamou was even reported to be financially and logistically supported by the French military, with whom his militia conducted counter-terrorist raids in the Menaka region until early 2019 (see Carayol 2018). While joint *Barkhane* – GATIA military operations *arguably* served counter-terrorist purposes, they *certainly* fulfilled Gamou's community's desires for score-settling against the Fulani, and have helped consolidate his local position of power.

These cases demonstrate that, in both Afghanistan and Mali, local strongmen adeptly create security knowledge with which to manipulate intervening counter-terrorist forces, serving extremely personal and parochial interests. While a convergence of instrumentalising practices occurs in both countries, the durability of Afghan strongmen's has nevertheless been more pronounced and long-term than that of their Malian counterparts. Afghan strongmen enjoy stronger and less-challenged autonomous links to their communities and to international interveners than those in northern Mali, who have never been able to exploit their relationships with international actors to the same degree and duration.

'Post-War' Governance Institutions

Interveners act as custodians of newly-formed (and more often than not very fragile) post-war governance orders. As shown above, by virtue of their international positions and aims to transform local political institutions, intervention activities produce globalised incentive structures that armed actors seek to exploit to better dominate their own political arenas, and potentially for personal enrichment. In Mali, like in Afghanistan, Disarmament, Demobilisation, and Reintegration (DDR) programmes – including efforts to integrate former combatants into the ranks of national security and defence forces – and other schemes associated with power-sharing agreements – such as the creation of new state institutions – translate into new opportunities for local elites to advance their financial positions and consolidate, extend, and create new patronage networks. The globalised incentive structures put in place create a common template from which local armed actors can garner important material and symbolic resources. However, the amount of external resources devoted to the two countries widely differ, and so do their distribution patterns. While Afghan armed groups have benefited from massive amounts of money and economic opportunities, the ones in northern Mali have had to work with significantly less and therefore had a much harder time expanding their patronage networks beyond immediate community lines.

The mode of intervention chosen in Afghanistan – the so-called 'Afghan model' of warfare (Biddle 2005-2006) – and the sudden infusion of resources that came with it, combined with the end of the Taliban regime, allowed former warlords to recapture their fiefdoms and take control of most state institutions, both locally and nationally. Hamid Karzai, selected as head of the

Executive during the 2001 Bonn conference, initially lacked the capacities to challenge the warlords militarily and hence chose to acknowledge their authority. Following the same logic, the U.S. administration adopted a 'warlord strategy' seeking to promote short-term stability and counter-terrorism operations over any long-term state-building endeavour and human rights concerns (Human Rights Watch 2002). In turn, the warlords took advantage of the new institutions to restructure their patronage networks and accumulate financial resources. Some, such as Ismail Khan, a warlord from western Afghanistan, became self-appointed (and later officially confirmed) governors[26] and took control of border posts, hence accumulating vast amounts of custom revenues; others, such as Mohammad Qasim Fahim, the leader of the Northern Alliance (the main anti-Taliban coalition) became ministers, securing extremely lucrative contracts with foreign forces. Fahim might in fact be the best example of such practices. His control of the Ministry of Defence gave him the ability to resist the international pressure to disarm his men while allowing him to recruit them into the Afghan National Army, hence keeping them on the state's payroll. He could in turn maintain one of his most prominent new sources of patronage, not least by controlling the DDR process and the benefit packages that came with it (Chesterman 2002, Giustozzi 2008). As Alex de Waal points out, in Afghanistan, 'the ruler [and the central state this ruler is supposed to represent are] only as strong as today's bargains with members of the political elite, who retain a great deal of autonomy' (2010, p. 41).

Since *Serval*'s objective was never regime change in Mali, northern armed groups have not benefited from the same types of largesse as Afghan strongmen, having to split the infusion of international post-war reconstruction funds with the Malian state. While the APR implementation has proceeded slowly, the associated creation of post-war institutional 'commissions' established to implement DDR, Security Sector Reform (SSR), Transitional Justice, and other peace-agreement institutions has nevertheless produced opportunities for personal enrichment for signatory armed group leaders and their selected representatives. Such positions provide relatively significant financial gains and other 'perks' that help consolidate patronage networks. These include guaranteed salaries for commission officials, per diems, hotel stays and meals in Bamako, chauffeurs, vehicles, and the occasional 'capacity-building' trip to Europe or North America.[27] Such financial resources and lifestyle changes remain politically-divisive amongst members of northern communities who regularly express their discontent over how little money reaches them, complaining that armed group leaders have 'betrayed their movement's cause in order to enjoy life in the capital.'[28] Nevertheless, as armed group leaders have dolled out positions within APR's institutional commissions to key allies or integrated loyal fighters into the security forces, they have retained a significant level of notoriety within their communities.

Likewise, by guaranteeing integration spots to members of influential families and loyal fighters, these leaders keep community criticisms at bay. As a result, APR signatories have inflated the number of their armed group's fighters in attempts to secure a larger portion counted in DDR, SSR, and other post-war governance institution activities, going so far as to enrol family members that were never involved in any fighting.[29]

These newly-organised APR-linked institutions clearly embody globalised incentive structures supported by the international community from which senior leaders of Malian armed groups consolidate their immediate patronage networks, and increase their personal financial gains. Compared to Afghanistan, however, the share of internationally-supported resources intended for Mali is small and unevenly split, which reduces the share destined for northern armed groups. As a result, the latter have a more difficult time expanding their patronage networks due to their comparatively meagre means to distribute amongst multiple clients. This also contributes to armed group fragmentation along community lines in Mali.

Conclusion

Enough with 'Sahelistan'! The oft-made comparison between the armed conflicts in Mali and Afghanistan is symptomatic of a broader tendency, among the news media and policy makers, to box together contemporary conflicts in places where a relatively 'weak state' is confronted with Islamic militantism, political violence, and deep societal divisions. As such, these conflicts have a lot in common. Yet, many of the commonalities of this important category of contemporary conflicts, we argue, are a reflection of global opportunity structures created through standard external military intervention enacted far beyond the cases of Afghanistan and Mali. In other words, it is not so much these conflicts' original patterns that are consonant, but rather the way interveners operate in these vaguely similar situations, and how non-state armed actors (and others) adapt to these newly-introduced opportunities. As Englebert and Tull point out (in the African context), where the international community sees a 'systemic breakdown and reconstruction as a new form of social contracting … [local actors] are more likely to view both as an opportunity to maximize their political fortunes. They may see reconstruction as the continuation of war and political competition for resources by new means' (2008, p. 118).

In this article, we show that, despite similar intervention strategies and convergence in armed group behaviour, the non-state armed actors operating in Afghanistan and Mali still adopt distinctive practices and techniques, which reflect how fundamentally different these two conflicts were in the first place. In both cases, foreign intervention has deeply altered patronage politics and incentivised domestic actors to instrumentalise international ones. Yet, we argue that largely different contexts – in terms of scale of war and intervention, state capacity, nature of the actors involved, and war economy – have led to significant variation in processes and outcomes. In Afghanistan, strongmen have gained considerable sway over their communities and a high degree of autonomy vis-a-vis the central state, while developing substantial relationships with foreign actors over decades of war. As a result, they have been able to take great advantage of newly-introduced opportunities after 2001. In contrast, their counterparts in northern Mali have had much more difficulty in exploiting their relationships with external interveners, and mobilising their communities beyond their immediate tribes and clans. While Afghan strongmen have utilised foreign resources to further their autonomy, in Mali, similar opportunities (though at a much smaller scale) have led to further fragmentation of the local political economy of violence since the international community's intervention began.

To be sure, comparing some of these conflicts' most fundamental features is not problematic per se. Yet, such endeavours must be handled with care. In pointing out the risks of amalgamation, we do not disregard the benefits of careful comparison. Instead, we aim to warn policy makers, pundits, the media, and others of the grave and concrete consequences of such discourses and associated practices when they are uncritically repeated and end up driving policy in war-affected countries. The casual conflation of these conflicts' patterns in their entirety produces a powerful grid of intelligibility that masks the idiosyncrasies and historical trajectories of these armed conflicts. It contributes to the production of approximate, even inaccurate, conflict knowledge that often serves as the basis for action and leads to the adoption of technical, cookie-cutter, solutions to complex, unique, political situations. The recipes that have failed elsewhere are adapted, 'improved', and rehashed in conflicts whose fundamental patterns broadly resemble those for which the original policies were designed.

Notes

1. In this article, all translations from French to English were made by the authors.
2. Interview with former French Defence official for Mali, Bamako, 2017.
3. On the method of focused, structured comparison, see George and Bennett (2005).
4. To remedy a general lack of contextual knowledge of theatres of intervention, governments and military forces have developed an array of educational initiatives and other mechanisms. These include courses to enhance their armed forces' cultural and linguistic awareness prior to deployment, hiring subject-matter experts, and developing various physical objects like portable 'smart-cards' with information on the operational environment, and field manuals. These programmes and methods, however, are widely viewed as extremely limited, insufficient, unhelpful, ad hoc, and often lead to misunderstandings and frustrations (Davis 2010, Fosher 2014, Holmes-Eber 2020).
5. The 'Sahelistan' label has been widely echoed by mainstream journalists and pundits – often with different meanings – since Fabius initially coined it. See, for

example, Légaré-Tremblay (2012), Heisbourg (2013, pp. 10–11), Laurent (2013), Leymarie (2013), Nougaryède (2013), Berger (2019), Kwasi *et al.* (2019).

6. On the role of brokers in civil war research, see, for instance, Baaz and Utas (2019).
7. On the importance of context when comparing civil wars, see Reno (2019).
8. *Barkhane* has a mandate to conduct counter-terrorism operations across five Sahelian countries, meaning that, of the 5,000 troops, not all operate in Mali.
9. For critical perspectives on the different classifications of states based on their governmental capacities, see Call (2011), Grimm (2014), Mazarr (2014).
10. In this, the Taliban differ from both Al-Qaeda and IS-KP.
11. JNIM was officially created in March 2017, bringing together fighters from *Al-Mourabitoun*, AQIM, and the more or less ethnically-distinct, geographically-anchored brigades of *Ansar Dine*. For more on these groups, see Raineri and Martini (2017) and Bencherif (2020).
12. Interview with Nigerien trafficker, Niamey, 2013; interview with Malian Arab businessman from Timbuktu, Bamako, 2013; interview with Malian trafficker from Kidal, Niamey, 2016.
13. On the role of opium in the Afghan conflict and state-building process, see Mansfield (2016).
14. Interview with Mauritanian security official, Dakar, 2016; interview with Malian Arab businessman from Timbuktu, Bamako, 2017; interview with Malian Arab businessman from Gao, 2017. See also International Crisis Group (2018).
15. Interview with two European security officials, Bamako, 2013. See also Marchal (2013).
16. Interview with three MINUSMA officials, Bamako, 2018.
17. On armed group fragmentation, see, for instance, Staniland (2014), Seymour *et al.* (2016), Woldemariam (2018).
18. On side-switching, see, for example, Christia (2012), Seymour (2014), Otto (2018).
19. In November 2016, Dostum was accused of publicly beating up a political rival, Ahmad Ishchi, before ordering his guards to abduct and sexually abuse him. Dostum, under investigation, flew to Turkey. He returned to Afghanistan over a year later, in July 2018.
20. On how such practices incentivise the escalation of violence by armed group leaders, see Tull and Mehler (2005).
21. Interview with MSA official, Bamako, 2017; interview with two Daoussahak notables, Bamako, 2018.
22. Interview with CJA official, Bamako, 2017.
23. Focus group with 14 residents of Gao, Bamako, 2013; interviews with three Arab businessmen from Timbuktu, Bamako, 2017. See also Chebli (2017).
24. Interview with Arab notables, Bamako, 2013.
25. Interview with Kel Ansar notable, Bamako, 2019; interview with MNLA supporter, Bamako, 2020.
26. On warlord-governors, see Mukhopadhyay (2014).
27. Fieldwork observations, Bamako, 2018 and 2019; personal communication with militia leader, 2019; meeting with HCUA official, Brussels, 2019.
28. Interviews with former MNLA and HCUA fighters, Bamako, 2017.
29. Interviews with DDR Commission officials and MINUSMA officials, Bamako, 2017.

Acknowledgments

The authors would like to thank the three anonymous reviewers and the editors of *Civil Wars* for detailed and constructive feedback, as well as Sean Lee for sound advice. The authors are also particularly grateful to all the participants of the "Politics of Comparing Armed Conflict" workshop, held on 13 October 2017 in Montréal, Canada.

Disclosure Statement

No potential conflict of interest was reported by the author(s).

Funding

This work was supported by the Käte Hamburger Kolleg, Centre for Global Cooperation Research [Post-Doctoral Research Fellowship]; Social Sciences and Humanities Research Council of Canada.

References

Ahmad, A., 2018. Why jihadist insurgencies persist. *OpenCanada.org*, 23 May.

Assemblée Nationale, 2013. Audition de l'amiral Édouard Guillaud sur les enseignements de l'opération Serval. *Commission de la défense nationale et des forces armées, compte rendu*, 74 , 22 May.

Autesserre, S., 2012. Dangerous tales: dominant narratives on the Congo and their unintended consequences. *African Affairs*, 111 (443), 202–222. doi:10.1093/afraf/adr080.

Autesserre, S., 2014. *Peaceland: conflict resolution and the everyday politics of international intervention*. New York: Cambridge University Press.

Ayad, C., 2017. Le Mali est notre Afghanistan. *Le Monde*, 19 Nov.

Baczko, A., 2016. Legal rule and tribal politics: the U.S. army and the Taliban in Afghanistan (2001-13). *Development and Change*, 47 (5), 1412–1433. doi:10.1111/dech.12276.

Baaz, M.E., and Utas, M., 2019. Exploring the backstage: methodological and ethical issues surrounding the role of research brokers in insecure zones. Civil Wars, 21 (2), 157-178. doi:10.1080/13698249.2019.1656357.

Bencherif, A., 2020. From resilience to fragmentation: Al Qaeda in the Islamic Maghreb and jihadist group modularity. *Terrorism and Political Violence*, 32 (1), 100–118. doi:10.1080/09546553.2017.1351956.

Berger, O., 2019. Face au péril d'un "Sahelistan". *La Voix du Nord*, 13 Dec.

Biddle, S.D., 2005-2006. Allies, airpower, and modern warfare: the Afghan model in Afghanistan and Iraq. *International Security*, 30 (3), 161–176. doi:10.1162/016228805775969555.

Bilgin, P. and Morton, A.D., 2002. Historicising representations of 'failed states': beyond the cold-war annexation of the social sciences? *Third World Quarterly*, 23 (1), 55–80. doi:10.1080/01436590220108172.

Bleck, J. and Michelitch, K., 2015. The 2012 crisis in Mali: ongoing empirical state failure. *African Affairs*, 114 (457), 598–623. doi:10.1093/afraf/adv038.

Bouhlel, F., 2020. *(Ne pas) Dialoguer avec les groupes "jihadistes" au Mali? Entre contradictions normatives, "manqués" de la médiation … et nécessité heuristique. Rapport d'étude de cas*. Berlin: Berghof Foundation.

Bourdieu, P., 1991. Description and prescription: the conditions of possibility and the limits of political effectiveness. *In*: J.B. Thompson, eds *Language & symbolic power*. Cambridge: Cambridge University Press, 127–136.

Call, C., 2011. Beyond the 'failed state': toward conceptual alternatives. *European Journal of International Relations*, 17 (2), 303–326. doi:10.1177/1354066109353137.

Carayol, R., 2018. A la frontière entre le Niger et le Mali, l'alliance coupable de l'armée française. *Médiapart*, 29 Nov.

Cecchinel, L., 2017. Le Mali est-il un nouvel Afghanistan? *l'ena hors les murs*, 467, 13–15.

Charbonneau, B. and Sandor, A., 2019. Power and comparative methods: performing the worlds of armed conflict. *Civil Wars*, 21 (4), 437–447. doi:10.1080/13698249.2019.1706144.

Charbonneau, B. and Sears, J., 2014. Fighting for the liberal peace in Mali? The limits of international military interventions. *Journal of Intervention and Statebuilding*, 8 (2–3), 192–213. doi:10.1080/17502977.2014.930221.

Chayes, S., 2006. *The punishment of virtue: inside Afghanistan after the Taliban*. New York: Penguin.

Chebli, D., 2017. L'échec de l'intervention française au Mali. *Libération*, 27 Jun.

Chesterman, S., 2002. Walking softly in Afghanistan: the future of UN state-building. *Survival*, 44 (3), 37–47. doi:10.1080/00396330212331343402.

Christia, F., 2012. *Alliance formation in civil wars*. New York: Cambridge University Press.

Craven-Matthew, C. and Englebert, P., 2018. A Potemkin state in the Sahel? The empirical and the fictional in Malian state reconstruction. *African Security*, 11 (1), 1–31. doi:10.1080/19392206.2017.1419634.

Davis, R., 2010. Cultural sensitivity in a military occupation: the U.S. Military in Iraq. *In*: J. D. Kelly, *et al.*, eds. *Anthropology and global counterinsurgency*. Chicago: University of Chicago Press, 297–310.

De Waal, A., 2010. Dollarised: Alex de Waal writes about the political marketplace. *London Review of Books*, 32 (12), 38–41.

Derksen, D., 2017. In Afghanistan, today's pro-government militias could be tomorrow's insurgents. *War on the Rocks*, 11 Dec.

Desgrais, N., Guichaoua, Y., and Lebovich, A., 2018. Unity is the exception: alliance formation and de-formation among armed actors in northern Mali. *Small Wars & Insurgencies*, 29 (4), 654–679. doi:10.1080/09592318.2018.1488403.

Dubois, O., 2018. Talataye: vivre dans la peur. *Journal du Mali*, 15 Feb.

Englebert, P. and Tull, D.M., 2008. Postconflict reconstruction in Africa: flawed ideas about failed states. *International Security*, 32 (4), 106–139. doi:10.1162/isec.2008.32.4.106.

Erforth, B., 2020. *Contemporary French security policy in Africa: on ideas and wars*. Basingstoke: Palgrave Macmillan.

Fabius, L., 2012a. *Entretien de M. Laurent Fabius, ministre des affaires étrangères, avec Europe 1 le 6 juillet 2012, sur la situation politique en Syrie, le retrait des troupes françaises d'Afghanistan et sur la présence du groupe islamiste AQMI au Mali*. Available from: https://www.vie-publique.fr/discours/185463-entretien-de-m-laurent-fabius-ministre-des-affaires-etrangeres-avec-e. Accessed 25 May, 2020.

Fabius, L., 2012b. *Déclaration de M. Laurent Fabius, ministre des affaires étrangères, en réponse à une question sur le terrorisme au Sahel, à l'Assemblée nationale le 3 octobre 2012*. Available from: https://www.vie-publique.fr/discours/186005-declaration-de-m-laurent-fabius-ministre-des-affaires-etrangeres-en-r. Accessed 25 May, 2020.

Fierke, K.M., 2007. *Critical approaches to international security*. Cambridge: Polity Press.

Fosher, K., 2014. Cautionary tales from the U.S. department of defense's pursuit of cultural expertise. *In*: R. Albro and B. Ivey, eds. *Cultural awareness in the military: developments and implications for future humanitarian cooperation*. Basingstoke: Palgrave Macmillan, 15–39.

Fragile States Index, 2020. Available from: https://fragilestatesindex.org. Accessed 2 Sep, 2020.

George, A.L. and Bennett, A., 2005. The method of focused, structured comparison. *In*: A.L. George and A. Bennett, eds. *Case studies and theory development in the social sciences*. Cambridge, MA: MIT Press, 67–72.

Giustozzi, A., 2008. Bureaucratic façade and political realities of disarmament and demobilisation in Afghanistan. *Conflict, Security & Development*, 8 (2), 169–192. doi:10.1080/14678800802095369.

Grimm, S., 2014. The European Union's ambiguous concept of 'state fragility'. *Third World Quarterly*, 35 (2), 252–267. doi:10.1080/01436597.2014.878130.

Hansen, L., 2006. *Security as practice: discourse analysis and the Bosnian War*. London: Routledge.

Heisbourg, F., 2013. A surprising little war: first lessons of Mali. *Survival*, 55 (2), 7–18. doi:10.1080/00396338.2013.784458.

Holmes-Eber, P., 2020. Lost in translation: anthropologists and Marines in Iraq and Afghanistan. *Small Wars & Insurgencies*, 31 (2), 340–358. doi:10.1080/09592318.2020.1713545.

Human Rights Watch, 2002. *Afghanistan's Bonn agreement one year later: a catalog of missed opportunities*. New York: Human Rights Watch.

International Crisis Group, 2018. Narcotrafic, violence et politique au nord du Mali. *Africa Report*, p. 267.

Jones, B.G., 2008. The global political economy of social crisis: towards a critique of the 'failed state' ideology. *Review of International Political Economy*, 15 (2), 180–205. doi:10.1080/09692290701869688.

Khong, Y.F., 1992. *Analogies at war: Korea, Munich, Dien Bien Phu, and the Vietnam decisions of 1965*. Princeton: Princeton University Press.

Kwasi, S., *et al.*, 2019. Prospects for the G5 Sahel countries to 2040. *ISS West Africa Report*, p. 25.

Lacher, W., 2013. Challenging the myth of the drug-terror nexus in the Sahel. *WACD Background Paper*, p. 4.

Lagneau, L., 2019. Général Lecointre: 'Nous avons été exagérément optimistes' après l'opération Serval. *Zone Militaire: opex360.com*, 26 Jul.

Laurent, S., 2013. *Sahelistan: de la Libye au Mali, au cœur du nouveau jihad*. Paris: Seuil.

Légaré-Tremblay, J.-F., 2012. Bienvenue au Sahelistan. *L'actualité*, 5 Sep.

Lema Landu, J.-J., 2018. Sahel: naissance d'un autre Afghanistan? *Ouest France*, 8 May.

Lewis, A., 2013. Why the Sahara Is Not the 'New Afghanistan'. *BBC News*, 6 Feb.

Leymarie, P., 2013. La guerre du "Sahelistan" aura-t-elle lieu? *Le Monde Diplomatique*, Jan.

Mac Ginty, R., 2012. Routine peace: technocracy and peacebuilding. *Cooperation and Conflict*, 47 (3), 287–308. doi:10.1177/0010836712444825.

Malejacq, R., 2010. L'influence des analyses théoriques de la conflictualité d'après-guerre froide dans la construction de la paix: l'exemple des conflits de la région du fleuve Mano. *In*: R. Bowd and A. Chikwanha, eds. *Understanding Africa's contemporary conflicts*. Pretoria: Institute for Security Studies, 3–24.

Malejacq, R., 2015. Warlords and the coalition in Afghanistan. *In*: G.A. Mattox and S. M. Grenier, eds. *Coalition challenges in Afghanistan: the politics of alliance*. Stanford, CA: Stanford University Press, 31–44.

Malejacq, R., 2019. *Warlord survival: the delusion of state building in Afghanistan*. Ithaca, NY: Cornell University Press.

Malejacq, R. and Mukhopadhyay, D., 2016. The tribal politics of field research: a reflection on power and partiality in 21st-century warzones. *Perspectives on Politics*, 14 (4), 1011–1028. doi:10.1017/S1537592716002899.

Mansfield, D., 2016. *A state built on sand: how opium undermined Afghanistan*. New York: Oxford University Press.

Marchal, R., 2013. Military (mis)adventures in Mali. *African Affairs*, 112 (448), 486–497. doi:10.1093/afraf/adt038.

Martin, M., 2014. *An intimate war: an oral history of the Helmand conflict*. New York: Oxford University Press.

Mazarr, M.J., 2014. The rise and fall of the failed-state paradigm: requiem for a decade of distraction. *Foreign Affairs*, 93 (1), 113–121.

McGreal, C., 2013. Clinton demands U.S. takes lead to combat 'jihadist threat' in North Africa. *The Guardian*, 23 Jan.

McNeill, C., 2017. "Playing the away game": AFRICOM in the Sahara-Sahel. *Political Geography*, 58 (1), 46–55. doi:10.1016/j.polgeo.2017.01.003.

Minear, L., Scott, C., and Weiss, T.G., 1996. *The news media, civil war & humanitarian action*. Boulder, CO: Lynne Rienner.

Mukhopadhyay, D., 2014. *Warlords, Strongman Governors, and the State in Afghanistan*. New York: Cambridge University Press.

Nougaryède, N., 2013. Le "Sahelistan" de François Hollande. *Le Monde*, 17 Jan.

Otto, S., 2018. The grass is always greener? Armed group side switching in civil wars. *Journal of Conflict Resolution*, 62 (7), 1459–1488. doi:10.1177/0022002717693047.

Paris, R., 2004. *At war's end: building peace after civil conflict*. Cambridge: Cambridge University Press.

Pérouse de Montclos, M.-A., 2015. Boko Haram and "Sahelistan" terrorism narratives: a historical perspective. *Afrique Contemporaine*, 255 (3), 21–39. doi:10.3917/afco.255.0021.

Peters, G., 2009. How opium profits the Taliban. *Peaceworks*, p. 62.

Pham, J.P., 2010. Emerging West African terror-drug nexus poses major security threat. *World Defense Review*, 28 Jan.

Pham, J.P., 2013. The uncanny parallels between Afghanistan and Africa. *New York Times*, 16 Jan.

Philp, C., 2020. Mali mission 'must learn Afghan lessons'. *Times*, 7 Mar.

Raineri, L. and Martini, A., 2017. ISIS and Al-Qaeda as strategies and political imaginaries in Africa: a comparison between Boko Haram and Al-Qaeda in the Islamic Maghreb. *Civil Wars*, 19 (4), 425–447. doi:10.1080/13698249.2017.1413226.

Reno, W., 1998. *Warlord politics and African states*. Boulder, CO: Lynne Rienner.

Reno, W., 2019. The importance of context when comparing civil wars. *Civil Wars*, 21 (4), 448–467. doi:10.1080/13698249.2019.1642614.

Reno, W. and Matisek, J., 2018. A new era of insurgent recruitment: have "new" civil wars changed the dynamic? *Civil Wars*, 20 (3), 358–378. doi:10.1080/13698249.2018.1497314.

Rubin, B., 2002. *The fragmentation of Afghanistan: state formation and collapse in the international system*. 2nd ed. New Haven, CT: Yale University Press.

Sandor, A. and Campana, A., 2019. Les groupes djihadistes au Mali, entre violence, recherche de légitimité et politiques locales. *Canadian Journal of African Studies*, 53 (3), 415–430.

Schmidle, N., 2009. The Saharan conundrum. *New York Times*, 13 Feb.

Seymour, L.J.M., 2014. Why factions switch sides in civil wars: rivalry, patronage, and realignment in Sudan. *International Security*, 39 (2), 92–131. doi:10.1162/ISEC_a_00179.

Seymour, L.J.M., Bakke, K., and Cunningham, K.G. 2016. E pluribus unum, ex uno plures: competition, violence, and fragmentation in ethnopolitical movements. *Journal of Peace Research*, 53 (1), 3–18. doi:10.1177/0022343315605571.

Staniland, P., 2014. *Networks of rebellion: explaining insurgent cohesion and collapse*. Ithaca, NY: Cornell University Press.

Suhrke, A., 2007. Reconstruction as modernisation: the 'post-conflict' project in Afghanistan. *Third World Quarterly*, 28 (7), 1291–1308. doi:10.1080/01436590701547053.

Suhrke, A., 2011. *When more is less: the international project in Afghanistan*. London: Hurst.

Tull, D. and Mehler, A., 2005. The hidden costs of power-sharing: reproducing insurgent violence in Africa. *African Affairs*, 104 (416), 375–398. doi:10.1093/afraf/adi034.

United Nations Office on Drugs and Crime, 2020. *World drug report 2020: 3. Drug supply*. Vienna: United Nations.

Wing, S., 2016. French intervention in mali: strategic alliances, long-term regional presence? *Small Wars & Insurgencies*, 27 (1), 59–80. doi:10.1080/09592318.2016.1123433.

Woldemariam, M., 2018. *Insurgent fragmentation in the Horn of Africa: rebellion and its discontents*. Cambridge: Cambridge University Press.

Woodward, S.L., 2017. *The ideology of failed states: why intervention fails*. Cambridge: Cambridge University Press.

Index